S0-ADT-330

B+5
8-08

DISCARDED
BERNARDS TOWNSHIP LIBRARY
32 S. MAPLE AVENUE
BASKING RIDGE, NJ 07920

DISCARDED

ALSO BY PETER PRINGLE

*Food, Inc.: Mendel to Monsanto—The Promises and Perils
of the Biotech Harvest*

Those Are Real Bullets: Bloody Sunday, Derry, 1972
(with Philip Jacobson)

Cornered: Big Tobacco at the Bar of Justice

S.I.O.P.: The Secret U.S. Plan for Nuclear War
(with William Arkin)

Chernobyl: The End of the Nuclear Dream
(with *The Observer,* London)

The Nuclear Barons (with James Spigelman)

Insight on Portugal: The Year of the Captains (with *The
Sunday Times,* London, Insight Team)

Insight on the Middle East War (with *The Sunday Times,*
London, Insight Team)

Day of the Dandelion (novel)

The Murder of

The Story of Stalin's Persecution of One of

Nikolai Vavilov

the Great Scientists of the Twentieth Century

Peter Pringle

Simon & Schuster
NEW YORK LONDON TORONTO SYDNEY

Simon & Schuster
1230 Avenue of the Americas
New York, NY 10020

Copyright © 2008 by Peter Pringle

All rights reserved, including the right to reproduce this book or
portions thereof in any form whatsoever. For information address
Simon & Schuster Subsidiary Rights Department,
1230 Avenue of the Americas, New York, NY 10020.

First Simon & Schuster hardcover edition May 2008

SIMON & SCHUSTER and colophon are registered trademarks
of Simon & Schuster, Inc.

For information about special discounts for bulk purchases,
please contact Simon & Schuster Special Sales at
1-800-456-6798 or business@simonandschuster.com.

Designed by Paul Dippolito

Manufactured in the United States of America

1   3   5   7   9   10   8   6   4   2

Title page photograph: Nikolai Vavilov in 1927 (N. I. Vavilov, *Five Continents*,
Rome: IPGR, 1997; orig. in Archives of the Russian Academy of Sciences,
ARAN, f. 803, op. 5, d. 1, l. 4).

Illustration credits are listed following the index.

Library of Congress Cataloging-in-Publication Data
Pringle, Peter.
The murder of Nikolai Vavilov : the story of Stalin's persecution of one of the
great scientists of the twentieth century / Peter Pringle.
p.   cm.
Includes bibliographical references and index.
1. Vavilov, N. I. (Nikolai Ivanovich), 1887–1943. 2. Plant breeders—Soviet Union—
Biography. 3. Plant geneticists—Soviet Union—Biography. 4. Stalin, Joseph, 1879–1953.
5. Political purges—Soviet Union—History—20th century. 6. Political atrocities—
Soviet Union—History—20th century. I. Title.
[DNLM: 1. Vavilov, N. I. (Nikolai Ivanovich), 1887–1943. 2. Stalin, Joseph, 1879–1953.
3. Genetics—Russia—Biography. 4. Communism—history—Russia. 5. Genetics—
history—Russia. 6. History, 20th Century—Russia—Biography. WZ 100 P9567m 2008]
SB63.V38P35 2008
630.92—dc22   [B]   2008003510
ISBN-13: 978-0-7432-6498-3
ISBN-10:   0-7432-6498-3

For Robert Haupt, 1947–1996,
for Natalia, Slava, Anton, and Andrei,
and, as always, for Eleanor

"You would have to have no love whatever for your country, you would have to be hostile to it, to shoot the pride of the nation—its concentrated knowledge, energy and talent! And wasn't it exactly the same . . . in the case of Nikolai Ivanovich Vavilov?"

—ALEKSANDR SOLZHENITSYN,
THE GULAG ARCHIPELAGO

Contents

Introduction

When I was a correspondent in Moscow in the last days of communism, I lived on a street named for Dmitry Ulyanov, Lenin's brother. Other streets nearby were a Who's Who of the old USSR and its socialist allies, even Ho Chi Minh. Many of the names meant nothing to me. Ulitsa Vavilova, Vavilov Street, was a mystery until one day a Russian friend told me the story of the Vavilov brothers.

The street had been named for Sergei Ivanovich Vavilov, a physicist of great renown. He became Stalin's president of the Academy of Sciences at the end of the Second World War and oversaw the beginnings of the Russian atomic bomb project. But it was Sergei's older brother, Nikolai, who was an even greater scientist and who was actually more famous, my friend said. Nikolai Vavilov was a botanist and geneticist, a plant breeder, an intrepid explorer, and an organizer of science. He had an ambitious plan to end famine throughout the world. He wanted to use the new science of genetics to breed varieties that would grow where none had survived before. The key was a treasure trove of genes he was sure he could find in the unknown and wild types that had been ignored by our ancestors as they started farming more than ten thousand years ago. To cultivate these crops, the early farmers selected the seeds of plants that looked strong and yielded more grain—visible characteristics. But

1

Vavilov was looking for the complex properties, such as the ability to withstand extremes of temperature and resistance to pests.

In the 1920s, Nikolai Vavilov roamed the world hunting for these wild varieties of wheat, corn, rye, and potatoes. He built the first international seed bank of food plants, a magnificent collection of hundreds of thousands of botanical specimens, a living library of the world's genetic diversity that would preserve species from extinction and could be used to breed his new miracle plants.

Nikolai's fame spread far beyond Russia, my friend told me. He was a leader of the biological world of the early twentieth century. His seed bank was the envy of his colleagues in Europe and America and they came to work with him at his plant breeding Institute in Leningrad.

In the first years after the 1917 revolution, Vladimir Ilyich Lenin understood the ultimate economic power of Nikolai Vavilov's dream—to push Russia into the forefront of world food production—and he supported Vavilov's expeditions. But Lenin died in 1924, and his successor, Josef Stalin, had a very different priority. Russians were starving. Stalin's forced collectivization of Russian agriculture had disrupted the harvests, and a widespread famine would claim millions of lives. The shortage of food was also a constant threat to the revolution.

Stalin gave Vavilov three years to produce his new miracle plants—an impossible task, as Vavilov knew. To breed improved varieties using the new science of genetics took ten to twelve years. Impatient and ruthless, Stalin charged the geneticists like Vavilov with treason, called them "wreckers" and "saboteurs." They were jailed or executed. Vavilov died of starvation in 1943 in jail. "Just imagine," said my Russian friend, "the man who wanted to feed the world died of hunger in Stalin's prisons."

For many years in the Soviet Union you couldn't read Vavilov's scientific papers or even mention his name, my friend continued. But after Stalin died in 1953, Vavilov was "rehabilitated"—pardoned—and his reputation as a great scientist restored. The street near my Moscow home was named for his brother, Sergei, but Nikolai Vavilov is the one who is remembered all over Russia today. He has many memorials and plaques where he lived in St. Petersburg, and where he died in prison in Saratov on the Volga.

"And so, there you have it," my Russian friend had concluded, "a Shakespearean tragedy about two brothers, two brilliant scientists caught in revolution, civil war, and Stalin's terror, where one is destroyed by the regime and the other becomes a tool of it."

The story of the Vavilov brothers, like so many other seductive Russian sagas, leaves the listener wondering how much is true, and how much folklore. What began for me as idle curiosity about a street's name turned into a long and fascinating path of discovery about the violent birth of genetics in Russia, a path that would also reveal an intimate portrait of a bourgeois Russian family trying desperately to survive revolution, civil war, and Stalin's terror.

Nikolai Ivanovich Vavilov was a *bogatyr*, as the Russians say, a man of incredible powers, a Hercules. He was indeed an international figure, a fearless explorer, a plant hunter who saw more varieties of food plants in their place of origin than any other botanist in his time. His collection of seeds from five continents captivated the scientific world.

In the early years of the genetic revolution, Vavilov changed the way scientists looked at their new bounty—the world's vast store of valuable plant genes. Now, in the age of biotech agriculture it seems obvious to us that if you want to create a better, sturdier variety of corn or wheat, you should explore the total genetic diversity of the botanical kingdom for those exotic genes. But back then, as scientists debated the practical use of Mendel's laws of heredity and the words "gene" and "genetics" had only just entered the vocabulary, Vavilov's concepts were radical and innovative.

Before biotechnology and even before Watson and Crick had broken the genetic code, Vavilov laid out a grand plan for "sculpting" plants to human needs, for synthesizing varieties unknown in nature. He opened the eyes of the world's plant hunters and breeders to new ways of applying their expertise, forcing them to think outside the limits of a single academic discipline—botany—to include geography, biochemistry, taxonomy, and archaeology. His contributions to pure science were not as profound as Darwin's or Mendel's; he did not expound a revolutionary theory, or new laws of nature, but in a more practical way his research would eventually contribute directly to the food supply of millions of human beings around the world. With his astonishing breadth of knowledge and outstanding capacity to organize a vast amount of material, he set the scene for the exploration and preservation of the earth's genetic resources—its biodiversity—not just in Russia but across the planet. He was one of the great scientists of the twentieth century.

People found Vavilov irresistible. As he scoured the world for exotic genes, the energetic Russian cut a dashing, impressive figure, far from the common image of the plant breeder in dungarees and soil-caked boots. He was a man of medium height, stocky and well proportioned. He was handsome, with brown eyes, heavy eyebrows, and a carefully clipped mustache. His dark hair was brushed straight back in a way that added to his incongruous elegance. Wherever he was, in the city or the jungle, he insisted on dressing like a tsarist professor, in a finely tailored three-piece, double-breasted dark gray suit, white collar and tie, and a felt fedora. In the tropics he exchanged the fedora for an imperial pith helmet. He was almost always cheerful, had a deep, Robesian voice and seemingly inexhaustible energy. He was agile, his walk light and fast, he worked all hours, needed little sleep, and could endure physical hardships for long periods, the perfect mettle for a plant hunter.

Like most young biologists of his day, Vavilov followed Mendel's laws of inheritance and bred his new plants accordingly. But he had a young, ambitious rival, a Russian peasant horticulturalist named Trofim Lysenko. Lysenko claimed, falsely as it turned out, that he could "train" plants by changing their environment, and that these changes would be inherited in the next generation. Lysenko promised Stalin he could meet the demand for new varieties of crops in three years, not a decade like Vavilov. For this and other reasons, Stalin supported Lysenko's work. Vavilov's battles with Lysenko resulted in what has been called the "biggest fraud in biology."[1] Certainly it was the most vicious anti-science campaign of the twentieth century. When Nikolai Vavilov was compelled to choose between Lysenko's anti-science speculations and the theory of genetics he knew to be true, he declared, "We shall go into the pyre, we shall burn, but we shall not retreat from our convictions."[2]

In the West, there have been several histories of science analyzing Lysenko's speculative claims, but Vavilov's scientific achievements, his exploration of the world's botanical diversity, his dream of ending famine, and the intimate and, in the end, tragic story of the Vavilov family in revolutionary Russia have barely been told. There have been many accounts of Stalin's prisons, but the official archives now available on Vavilov's arrest and interrogation are among the most complete from that era.

Vavilov's son, Yuri, an eighty-year-old physicist living in Moscow, has published selected texts in Russian of his father's secret police dossier—a classified file that he was given as the only direct relative of a pardoned victim of Stalin's terror. Yuri Vavilov is also the keeper of the family archive, the custodian of boxes of scientific papers, as well as letters and photographs that amazingly survived his father's arrest.

Relatives, friends, and colleagues, including some who courageously hid papers from the secret police, have added invaluable reminiscences that give details of Vavilov's life beyond his scientific contributions. The original notes and reports of his expeditions were destroyed by Stalin's agents, but most of a manuscript he was writing at the time of his arrest survived. It was to be called "Five Continents," a chronicle of his plant hunting expeditions. Remarkably, eight volumes of his official letters survived the nine-hundred-day German army siege of Leningrad and were found, after the war, in the basement of his Institute. The breadth of this archive reflects an astonishing life, especially for one who could not live it to a natural end.

Yuri Vavilov has carried the burden of his father's murder with much the same stoicism that his father summoned during his stand against Stalin. At our meetings in Moscow and St. Petersburg, Yuri produced his father's legacy, piece by piece, including Nikolai's love letters to Yelena Barulina, Yuri's mother. The one journey we were unable to take was to his father's final resting place. Nikolai's body was dumped, with the bodies of other prisoners, in an unmarked mass grave.

A decade after my conversation about the street named Ulitsa Vavilova, I boarded a train in Moscow bound for the Vavilov ancestral home, the village of Ivashkovo, about sixty miles west of Moscow along the old trade route to the Baltic Sea. Today, it is a typical northern Russian village, with its rolling farmland and birch forests, its wooden cottages with intricately carved shutters, a shop, a school, and a crude concrete war memorial to those lost in the Great Patriotic War of 1941–45. A whitewashed Orthodox church dominates the village, its dome and golden onions having survived weather and conflicts and even seventy years of communism.

In midsummer, when the trees are in full leaf and the village paths are overgrown with wildflowers, Ivashkovo appears to be a charming, even peaceful place. But like many rural communities in Russia, the pastoral beauty hides a violent, wrenching history. The families who survived were tough, resolute people who had learned how to endure incredible hardships.

In the sixteenth century the land and its serfs were owned by aristocratic and greedy *boyars*; in the seventeenth century the village was pilfered by Polish troops. The villagers—freed like other serfs across Russia in 1861—enjoyed a brief respite before the Great Famine of 1891 that claimed thousands of lives. A few short years of relative prosperity were followed by the 1917 revolution. The forced collectivization of the farms in the early 1930s brought chaos and more famine, and Stalin's terror touched even small Russian villages across the countryside. Ivashkovo's church, the center of village life, was closed and converted into a storage barn.

In 1941, when the Germans stormed into the village, it had about two hundred houses. Only forty remained when the Germans left at the beginning of 1943. They carried out executions and beatings, and all able-bodied males were sent to work in Germany. On the last day of the occupation, the Germans killed a twenty-one-year-old youth named Leosha because they thought he was a partisan. He had been drafted into the Red Army, but was too sick to go. They beat the youth to death in the square in front of the villagers, including women and children.

The cruelty did not end with the occupation. The Orthodox monk who was in charge of church had persuaded the Germans to reopen it, and each Sunday it was filled with worshippers. When the Germans retreated and the Red Army returned to the village, the monk was accused of collaborating with the fascists and executed along with the church warden.

For all the recurring tragedy, the villagers brighten when a visitor wants to learn more about Nikolai Ivanovich, Ivashkovo's most famous son. In the village school, photos of the Vavilov family are on permanent display: Nikolai Ivanovich's father, Ivan Ilyich, an upright, somewhat stern figure who went to Moscow and made his fortune in the textile industry; his mother, Alexandra Mikhailovna, a matronly babushka in a black coat and a black scarf tied tightly around her head; his younger brother Sergei, shown in a photo taken before the revolution, proudly displaying his tsarist military uniform. At the center of this village shrine is a photo of Nikolai Ivanovich himself, a, handsome, intense-looking young man with dark, piercing eyes. In 1943, when the villagers discovered he had died in prison, they wanted to march on the Kremlin, but they were told they would be arrested.

The Vavilov family chronicle moves rapidly from rural serfdom in Ivashkovo to urban wealth, through revolution and civil war. So much was going on all the time in Nikolai Ivanovich's crowded life. There he was with his adoring students in the potato fields of Saratov, or in his Leningrad office in an abandoned tsarist palace with gilded ceilings and crystal chandeliers and maps of the world strewn over the floor. There he was in his three-piece suit and his fedora on plant hunting expeditions in the peaks of the Pamirs, in Afghanistan, Ethiopia, Mexico, or the jungles of Bolivia, facing death from wild beasts or armed bandits. There he was breezing through the laboratories of the icons of genetics in England, France, Germany, or the United States; or dominating government conferences in Moscow and Leningrad, defending genetics.

Wherever you enter Nikolai Ivanovich's life you will be swept along by the very pace at which he lived. "Life is short," he would say, "we must hurry."

Ukraine, August 6, 1940

The black sedan, a Soviet version of the American Ford, hurtled along a dirt road from Chernovtsy spreading clouds of dust over the ripening wheat fields. Inside the car were four men dressed like government officials in dark suits and ill-fitting fedoras.

As the road started to climb into the Carpathians near the border with Romania, the men met another car coming down the hill toward them. The car was limping along with a puncture, but when the black sedan stopped it was not to offer help.

"Where is Academician Vavilov?" one of the four men shouted from the car window. "We must find Academician Vavilov."[1]

In the second car was a young botanist, Vadim Lekhnovich, a member of a Commissariat of Agriculture expedition led by the Soviet Union's chief geneticist and plant breeder, Nikolai Ivanovich Vavilov. It was August 6, 1940. Europe was in flames, the Battle of Britain was engaged, but western Ukraine was basking peacefully in the summer sun. The botanists had been in the fields looking for rare specimens of wild grasses that could be bred into new forms of wheat able to withstand the inhospitable climates of the northern steppes.

For Lekhnovich, the intensity of the men in the black sedan, even the rude one who was shouting, had broken into their peaceful pursuit of plant hunting, but the urgent request for Nikolai Ivanovich did not seem out of the ordinary. Vavilov was an important scientist who was frequently summoned to Moscow at short notice.

"Nikolai Ivanovich is with the others, collecting specimens," Lekhnovich called back. "Is there an emergency?"

The man in the black sedan glared and spat out an answer.

"Academician Vavilov has important official documents about grain exports. They are needed immediately at the Commissariat of Agriculture."

The cold, demanding voice was suddenly unsettling. This was no idle bureaucrat.

"Where is Academician Vavilov?" the man demanded again. "Tell us where we can find him."

"He is with the others, in a field farther up the mountain—" Lekhnovich began, but before he could finish, the black sedan accelerated away, the dust billowing.

Lekhnovich coaxed his crippled vehicle back down the mountain to Chernovtsy and the university hostel where they were all staying.

At dusk, Nikolai Ivanovich returned with his botanists to the hostel. The four men in the black sedan were waiting for him. As he got out of his car, the door of the black sedan opened, and one of the men jumped out. He began talking earnestly with Nikolai Ivanovich, who then got into the sedan and it drove off. The guard at the hostel, who had overheard the conversation, reported to the botanists that the men told Nikolai Ivanovich he was needed urgently in Moscow. He had gone with them, saying that he would return.

Shortly before midnight, two of the four men returned to the Chernovtsy hostel. They carried a note for Lekhnovich from Vavilov, penned in his own distinctive handwriting.

"In view of my sudden recall to Moscow, hand over all my things to the bearer of this note. N. Vavilov, August 6, 1940, 2315 hours."

The two men insisted, politely but firmly, that all Vavilov's belongings should be put into his suitcase, not leaving anything out, not even a scrap of paper. They said that Vavilov was already at the airport and was waiting for his belongings before flying to Moscow.

Lekhnovich and another of the botanists, Fatikh Bakhteyev, did as they were told. As they packed the papers, even scraps of Vavilov's notes, they wondered why Nikolai Ivanovich had not been given a chance to pack his own bag, or, more importantly, to give instructions to the staff on how to continue the expedition in his absence. They decided that one of them should accompany the bags to the airport to get the orders directly.

Bakhteyev volunteered to go. They took the luggage out to the car where the men were waiting, one of them already at the wheel. Bakhteyev started to explain why he had to go with them and began to get into the car. But as he opened the door, one of the men forced Bakhteyev out of the way, pushed him to the ground, and jumped into the sedan as it drove off.

Nikolai Ivanovich Vavilov had disappeared into Stalin's prisons.

CHAPTER 1

Moscow, December 1905

The first Russian revolution began in January 1905 when tsarist guards at the Winter Palace in St. Petersburg fired on a peaceful demonstration demanding an end to the monarchy. One hundred and thirty were killed. The massacre led to workers' strikes in major cities and to peasant revolts in half of the provinces of European Russia. The tsar introduced reforms but the protests continued, ending in bloody fighting in Moscow. When the tsarist troops entered the workers' enclave of Presnya, a new industrial suburb on the outskirts of Moscow, several hundred were killed and factories and houses destroyed by artillery fire. Ivan Ilyich Vavilov, his wife, Alexandra Mikhailovna, and their family lived in a wooden house with an apple orchard at No. 13 Middle Presnya.

The sound of gunfire was closer now and not just the pop, pop of Mauser revolvers that had been heard all week and sounded like fire-crackers, but the sharp report of rifles, echoing across the ponds and along the low marshy banks of the frozen River Presnya. On the porch outside the kitchen of No. 13 Middle Presnya stood Alexandra Mikhailovna, bundled up in her gray woolen coat. She was waiting anxiously for her elder son Nikolai, then eighteen, to return from high school.[1] Through the gentle snow and the gathering dusk she could make out slender figures in groups of two and three, carrying sacks and rifles, or maybe they were staves and shovels, she could not be sure. They were ghostlike figures moving swiftly in and out of the shadows on the far side of the street, never stopping long enough in one place to form the outline of a human, always keeping to the path and dodging their booby trap wires strung across the street to catch the gendarmes. During several days of the uprising, Presnya had become isolated from the rest of the city, cut off by homemade barricades. A proletarian encampment, the strikers called it.

The night before, the Presnya *druzhinniki*, the armed revolution-aries, had captured six of the tsarist artillerymen, brought them to a factory, lectured them about the need for a revolution, and then let them go.[2] The workers wanted a dialogue with the troops, but everyone in Presnya knew that sooner or later the government troops would attack.

Alexandra Mikhailovna's husband, Ivan Ilych, was a director of a trading company that sold the products of Presnya's largest textile mill, the Trekhgornaya Manufaktura, or Three Hills Manufacturing Company, founded in 1799.[3] Before going to work that morning he had told the family that the attack would probably not be on that day. The feared Semyonovsky Regiment, the "pacification troops" as they were called, had not yet arrived from St. Petersburg, and the local Moscow garrison did not have enough reliable troops to be certain of success. There was even a possibility that the reinforcements might be delayed several days.

Workers from the textile mill had also joined the uprising and production at the Three Hills factory was at a standstill. The police had joined the *druzhinniki*, and even the Cossacks, the tsar's most loyal force, had disobeyed orders to disperse rioters.

During the day, Alexandra Mikhailovna had heard that the artillery guns had been hauled into place ready for the attack. She knew that her boys were unlikely to be home on time, especially Nikolai. He was too curious for his own good, always getting involved. Only the day before the boys had helped build barricades, using some of the new wooden fencing from the orchard. Her younger son, Sergei, aged fourteen, had returned from school, but Nikolai was still out there, somewhere. Sergei had lost sight of his brother at the Gorbaty Bridge.

Alexandra Mikhailovna tried to be calm. Nikolai had only gone to see the barricades, she told herself, and he was big enough to look after himself. Nikolai was the stronger of the two boys; he was quick to use his fists in a street brawl. But they were using rifles and grenades out there.

As Alexandra Mikhailovna scanned the riverbanks, a grenade exploded harmlessly in the marsh. Another grenade followed and then a rocket arched up into the sky. It seemed to come from the area of Kudrinskaya Square and burst in midair, pieces of shrapnel clattering onto the roof of the house next door, and onto the kitchen porch. Out of the smoke came a figure running toward the house. The shooting started again and the figure disappeared, and then reemerged, darting, first this way, then that. Alexandra Mikhailovna rushed onto the porch calling out for Nikolai. Then she saw him and pulled him inside.

The attack on Presnya came in all its fury on December 17. A remorseless artillery barrage started before dawn and continued for fourteen hours, and by the morning of the 18th, the resistance had crumbled. The tsar's troops entered the enclave and quickly cleared the barricades. Leon Trotsky, then a twenty-six-year-old revolutionary leader, would later call the uprising "a majestic prologue to the revolutionary drama of 1917."[4]

Ivan Ilyich, like most of the new entrepreneurs who had risen from the peasantry during the rapid industrialization of the country, understood that the days of tsarist rule were numbered. As a factory director, he had seen firsthand the "revolutionary rise of the proletariat."[5] While he prayed for a peaceful transition to some kind of democracy, he expected the tsar to be overthrown. He expected to lose his fortune, the comfortable life he had built for himself and his family, and the three houses he owned. His plan to send Nikolai and Sergei into the textile business seemed doomed.

Ivan Ilyich was a realist. The uprising would be followed by reprisals even harsher than before, he believed, and that meant renewed hatred toward the monarchy. Russians with money, fearing total revolution, were already leaving. Foreign investment, British, French, and German money that had helped businesses like his, would soon dry up. Ivan Ilyich was a wealthy and highly respected resident of Presnya and would also become a member of the Moscow City Council. Prior to the 1905 uprising, Ivan Ilyich had bought land, including an orchard, in Middle Presnya where he would build three houses and four outbuildings. When the revolution came, he feared that he would not only lose his land and his fortune, he would be forced into exile.

With such thoughts, a weaker man, a man less anchored by religious faith and less committed to his family responsibilities, might have settled into deep despair, or moved quickly abroad. But Ivan Ilyich was also a patriot. He believed that Russia deserved a new order and he was prepared to do his part to bring it about.

For the next several months, Presnya was quiet but not peaceful. Christmas was miserable; the streets were filled with anxious citizens, many trying to decide whether to stay or leave the country. Alexandra Mikhailovna tried to run the Vavilov household as though nothing was happening. Ivan Ilyich insisted on going to the factory, and he did not encourage discussion of the siege at the dinner table. His exception was always a plea for mercy for the victims of the uprising in the family prayers that started and ended each day.

As part of her own effort to impose normality, Alexandra Mikhailovna held a large party for Nikolai's angel day, commemorating St. Nicholas, and the guests played the usual charades and other games. The schools were closed, and she tried not to let the boys out of the house on their own. When they did slip around her, Alexandra Mikhailovna watched from the kitchen window. Sometimes when the boys disappeared from view, she would come out onto the porch and call them home. Anyone seeing her there, outside the door normally used by tradesmen, might have mistaken this prominent matron for a servant. She was dark-skinned and always dressed in black with a black head scarf tied tightly around her head like a cleaning woman.

Life had not always been so comfortable. Alexandra Mikhailovna was the daughter of Mikhail Asonovich Postnikov, an artist who was employed as a fabric designer in the textile factory. When she was sixteen, her father brought Ivan Ilyich home from work, and the two had fallen instantly in love. He was attracted by her big wide eyes and her kind face, and she by his lean good looks, his uprightness, his godliness, his strength of character, and his self-confidence.

In those days, Ivan Ilyich had been working for the textile factory for only a few years and his future success was by no means assured. Alexandra Mikhailovna had especially liked the way he took care of his appearance even then. It was clear to her, indeed to anyone who met Ivan Ilyich, that he was determined to improve himself.

Ivan's father, Ilya, had been an indentured servant whose life was owned and controlled by the Streshnev family. They had bought the village of Ivashkovo, where the Vavilovs came from, in 1668. Like all serfs, Ilya could not leave the village or marry without permission, and could even be forced to marry against his will. He could be flogged, or sold to another master and separated from his family forever. The Emancipation Act of 1861, passed when Ivan was

two years old, freed the serfs and the Ivashkovo community blos-
somed, like others across Russia. The villagers opened a post office
and started a credit company and sought markets for their produce
in Moscow.[6] They grew vegetables for their own consumption but
the cash crop was flax and in summer the surrounding fields were
covered with a blanket of soft blue flowers. At harvest time, Ivan
and his brother Ilya would help collect the flax seed, separate the
silky fibers from the reedlike stems, and prepare them for sale to
the Moscow linen factories. They earned pocket money from selling oddments like rawhide, bristle, and cat pelts. That peaceful life
ended when the boys' father died suddenly on a business trip to St.
Petersburg and the family lost its breadwinner.[7]

For a youth in those days, an assured route out of Ivashkovo was
to become an apprentice indentured to a Moscow factory. Much as
Ivan and his brother dreamed of taking part in Russia's industrial
revolution and joining the rural migration to the cities, they knew it
was a perilous journey. The new suburbs were overcrowded, there
was rising crime, no sanitation, and the risk of disease. Sometimes it
seemed better to stay behind and become a part of the new, vibrant
village life. But Ivan was given as a *malchik*, an errand boy, to the
Moscow merchant Saprikin, who traded in manufactured goods and
lived in Presnya.[8]

Ivan Ilyich had a strong baritone voice and at the age of ten was
accepted into the choir of the Orthodox church on the estate of
Princess Nevitskaya in the growing industrial sector of southwest
Moscow that included Presnya. He could study under the protection of the church, but he found his duties tedious and restrict-
ing and preferred his work as an errand boy. Soon, he was a shop
assistant in a textile store. His hard work and his organization skills
were quickly rewarded and he rose rapidly into the highest ranks
of the firm.

Ivan Ilyich and Alexandra married on January 8, 1884, in the
same Orthodox church of St. Nicholas where he had sung in the

choir. Ivan was twenty-one, his bride was eighteen. As evidence of Ivan's new social standing, printed invitations invited guests to celebrate the wedding with a dinner and an evening ball in the nearby royal village of Kudrino, the home of Princess Nevitskaya.

Alexandra Mikhailovna would bear seven children. Three died in childhood, one of them, a boy, Ilya, at the age of seven. She never spoke about it, not even to the family. She just said he had been short-lived, "like a fragile plant."[9] The other children included two daughters, Alexandra and Lydia, and two sons, Nikolai and Sergei. Their birth certificates registered the children as "Moscow bourgeois" and they were brought up under simple but rigorously enforced rules of modesty, temperance, hard work, and self-discipline. When the boys strayed from this strictly imposed orderliness, Ivan Ilyich did not hesitate to use the strap. Sergei, being the more pliant of the two, would take his punishment meekly, without protest. Nikolai would defy his father's right to beat him, once climbing onto the sill of a second-story window and threatening to jump unless his father backed down. On that one occasion, the threat worked.[10] Despite her access to wealth, Alexandra Mikhailovna had furnished the Vavilov home with necessities only, the furniture was functional, not overdone, and on the walls hung reproductions of classical paintings.

Although Alexandra Mikhailovna ran the home, the head of the family was, without question, Ivan Ilyich. She accorded him the respect and loyalty that the couple considered proper. A dutiful wife, she referred to him in the third person, announcing to guests seeking his presence that "Himself is walking in the garden." Ivan Ilyich insisted that the boys should address their parents formally as "Mother" and "Father" and that, as children, they should be addressed by their first names, without a patronymic.[11]

Ivan Ilyich was indeed a powerful figure in Presnya in the new merchant class that inhabited the outer environs of Moscow, known as "calico city" for its profusion of textile mills. He was a clever man, self-educated, like his co-managers in the new emerging Russian bourgeoisie. He traveled widely inside Russia wherever the Three Hills Company sold its goods, including St. Petersburg, Riga, Odessa, Bukhara, Samarkand, and Tashkent. He had his own pavilion at the Nizhny-Novgorod fair. Preoccupied with business, he had little time for his family. Yet unlike some of his colleagues he was a liberal and a humanitarian, especially when he was, as he used to say, "out in society."[12] While he did not support the extreme goals of the budding revolutionaries on the factory floor, he believed in fair wages for a good day's work and proper working and living conditions. Barrack-style accommodation provided for workers at the Three Hills factory was reckoned to be among the best, with rows of up to three hundred beds in an open warehouse, most of them shared by workers on different shifts. There were also barracks for married couples and for families, with eight people to a room.[13] Indeed, Three Hills had a reputation for looking after its workers. In 1900, at the World Exhibition in Paris, the company had won the Grand Prix and three gold medals—two of them "For Efforts in Making Workers' Life Comfortable" and "For Training Apprentices."[14]

Despite his foreboding about Russia's future and the obvious desire of both Nikolai and Sergei to study science, Ivan Ilyich wished his sons to succeed him in business, and instead of sending them to the academic gymnasium he sent them to the Emperor's High School of Commerce in Moscow. As he saw it, the sciences were good fields for women. His two daughters, Alexandra and Lydia, would become medical doctors. Men went into business.[15]

At the emperor's high school, discipline was enforced by so-called uncles—retired soldiers who sat on duty in the corridors, the cafeteria, and washrooms. French and German were taught by native speakers, and science was also high on the curriculum.[16] Ivan

Ilyich knew that turning the boys toward commerce would be an uphill struggle. From an early age, both Nikolai and Sergei were fascinated by science. Nikolai collected plants, created his own herbarium, and liked to play with frogs in the Presnya ponds. One of his favorite pastimes in winter was to try and find out if the summer creatures were alive during their winter anabiosis at Gusev Island along the Presnya River.

Sergei often joined him in chemistry experiments, one of them ending disastrously. One day Nikolai had learned how to produce ozone at school and he brought home the chemical ingredients. He poured sulfuric acid over a mixture of potassium permanganate and the mixture exploded, harming the sight in his left eye. The doctor was called from the factory clinic, but he could do nothing and Nikolai had an eye disorder, a slight blurring of the vision in that eye, for the rest of his life.[17]

Still trying to steer his sons toward a career in business, Ivan Ilyich hired a tutor to lecture them on such topics as the "Honor of Commerce and Industry and Its Usefulness to Society," and the "History of Commerce from the Phoenicians to the Present." But Nikolai, in particular, was not impressed. He had already decided that he wanted to be a biologist. Sergei followed his lead into science and took up physics.

Nikolai Ivanovich's first choice of a career was medicine; he wanted to go to one of Russia's ten universities and become a doctor like his sisters, but the emperor's high school of commerce did not teach Latin, a required language for university entrance. He was impatient to begin his higher education and rather than take another year to learn Latin, he decided to study agriculture. He entered the Petrovskaya Agricultural Academy, a magnificent two-story, neoclassical palace in richly landscaped gardens on the outskirts of Moscow.

CHAPTER 2

The Petrovka and Katya

In 1900, biologists rediscovered Mendel's laws of hered-ity, first postulated in 1865. Mendel's theory of particles of heredity—later to be called genes—stored in the reproductive cells had been ignored for thirty-five years. But then at the turn of the century scientists confirmed his theories; his paper was "rediscovered" and the new science of genetics was born in Europe and America. Academic institutions began teach-ing new Mendelian breeding techniques for plants and ani-mals. In Moscow, the premier college for such studies was the Petrovskaya Agricultural Academy, known affectionately as the "Petrovka." In the fall of 1906, Nikolai Ivanovich began his studies there.

Nikolai Ivanovich took the Petrovka by storm, quickly establishing a reputation as one of its brightest students.

"There goes Vavilov," they would whisper as the freshman hur-ried by, weighed down with books and oblivious to the attention of his new admirers.

"See how he ate his ice cream before his soup," they chuckled, as they watched him rush absentmindedly through his lunch in the cafeteria, anxious to return to his studies.

During an official visit of the tsarist Ministry of Agriculture, which ran the Petrovka, the young student was holding forth on the subject of plant breeding when a small green lizard crawled out of his pocket and began climbing up his jacket. Much to the amusement of his audience, Nikolai Ivanovich calmly picked up the lizard, folded it into his handkerchief, and put it back into his pocket.[1]

The Petrovka was one of eighty institutions of higher education throughout Russia.[2] Teaching at the Petrovka was intense. There were no vacations—classes began in September and ended without a break in July, followed by two months of fieldwork on a farm or at an experimental station. For the beginning of his four-year course, Nikolai Ivanovich vowed to "study and firmly learn"—to master—the curriculum in his first two years. "Let's wait before we look into the future. Let's pause in the present," he wrote. Only after the course had been mastered would he allow himself to "move towards bright openings and more joy."[3]

His diary for these years reveals a young man driven to succeed, but also open to the possibility of failure. "Do what you can," he wrote to himself. "If you can't do something you wanted to do, then you will be forgiven, but if you don't want to try to do anything, you will not be forgiven." Even in those early days science was his only desire. "I passionately want science. I love it. It's the purpose of life. Only in science can one experience enthusiasm." And science should be used to improve life on earth. It was important "not to engage in utopianism, but to engage in everything that brings joy, calmness of emotion and reason." Several times he urged himself to work harder, not "to spread myself too thin . . . to concentrate on something."

He became famous among the students for working long hours. On field trips Nikolai Ivanovich would stay working as long as it was light, then he would sleep with the laborers in their barracks and be up again before dawn.[4]

He impressed his teachers and especially his professor, Dmitry Pryanishnikov, a world authority on soil science. Pryanishnikov was twenty-two years older than Nikolai Ivanovich, but he began treating his student as an equal, declaring to anyone who would listen that this young Vavilov was a genius. "And we do not call him this only because he is our contemporary," said Pryanishnikov.[5]

The Petrovka was part of the reforms that the tsar hoped would discourage future uprisings by producing graduates to help the peasants improve the country's agriculture.[6] The Petrovka's professors and students were enthusiastic about this task to "enlighten" peasant farmers and passionate in their belief that they, as agronomists, were making as worthy and honorable a contribution to society as the more glamorous physicists and chemists. Nikolai Ivanovich would write in his diary that he was ready to "commit his life to understanding nature for the betterment of humankind" and he pledged to "work for the benefit of the poor, the enslaved class of my country, to raise their level of knowledge."[7]

But the challenges were enormous. Russian grain yields were one third of the levels in France and Germany. The academy produced graduates trained in new farming methods, such as crop rotation and breeding, but the government basically ignored them. Russian farms were performing well enough—for the needs of the aristocracy and upper class. Russia was the second largest exporter of grain after America and the large estates were providing the best-quality grain for the granaries of Europe. The revenue supported the rich and landed. The peasants continued with their medieval agriculture, including wooden plows, and they barely had enough to eat. Local community organizations, a few private landlords, and private agricultural societies launched their own improvements in an attempt at "enlightenment." They imported new farming techniques from Europe and America, but their funds were limited.

Most landlords were stubbornly ignorant of advances in agricultural science, and the peasant underclass—90 percent of the population—resisted replacing methods that had been used for generations. The Petrovka students had tried to bring changes and the academy had been regarded as a "revolutionary hotbed" that even favored admitting female students.[8] The government fired what it regarded as the most troublesome professors, no new students were accepted, and the Petrovka was to be closed. But in 1891, the thirtieth anniversary of emancipation, came the Great Famine. Like others in earlier times this famine was brought on by drought.

While most people think of Russian farming as a battle against the cold, the most fertile lands along the Volga, which flows from north of Moscow 2,293 miles into the Caspian Sea, often suffer from blistering, cloudless skies that kill crops, animals, and people. In the summer of 1891–92 a prolonged drought in the Volga basin destroyed the harvest, and several hundred thousand people died of starvation. The agricultural authorities could no longer ignore the need to modernize, but their changes were largely cosmetic. The academy accepted students again, but entrance was restricted: no Jews were admitted and student fees had to be paid in advance, an effort to keep out radicals. The academy was also given a new name, the Moscow Agricultural Institute, but it was still known to students and staff as the Petrovka (as it will be referred to throughout this book). A new minister of agriculture with a science background was appointed and funds were made available for plant breeding stations, on the American pattern. It would not be enough. In the 1905 uprising, peasants tried to seize the large estates. The rebellion was crushed, a new package of reforms granted, and agricultural production started to improve. Still, Russian agriculture stumbled into the twentieth century far behind Europe and America. Reflecting the radicalism of the Petrovka, Vavilov wrote in his diary about "issues that ought to be revised."[9] They included, "religion, family life, marriage, attitude to women, women's issues, the sex question, upbringing, school."

•   •   •

It was also a time of great intellectual excitement in biology—especially for crop breeding. Darwin had left biologists with a puzzle. He had not explained the mystery of inheritance. How were the adaptations that he said were the cause of evolution passed on from one generation to the next?

Darwin had suggested that there might be two types of inheritance: "soft" and "hard." The soft inheritance theory suggested that organisms would pick up adaptations during their lifetime from the environment, and these adaptations would somehow change the constitution of a plant and would be inherited. The hard inheritance theory suggested a fixed set of factors in the organism that was passed to the offspring generally unaffected by the environment.

The idea of the inheritance of characteristics acquired during an organism's lifetime was first suggested by the French botanist Lamarck. The idea of a fixed set of factors had been put forward by Mendel in 1866, but his work had been ignored—until it was confirmed by three European scientists in 1900. Mendel, crossbreeding peas, had shown that certain characteristics, the color of the flowers or the shape of the seeds, reappear in subsequent generations according to a definite pattern. When he crossed purple flowers with white the offspring of the first generation were all purple. But when he allowed the purple flowers from the cross to fertilize themselves, they produced three purple flowers for one white. He concluded that there was no mixing of the colors, but that one color, purple, was "dominant" and one color, white, was "recessive." In succeeding generations both dominant, purple, and recessive, white, appear, showing no mixing of the characteristics. The factors, later to be called genes, apparently remain unchanged. This idea of unchanged genes meant that plant breeders could look for the hidden genes— resistance to disease, for example—and expect the gene for disease resistance to appear at some time in later generations.

The rediscovery of Mendel's work caused a revolution in biology but especially for breeders of animals or plants. Plant breeders needed to know how much Mendel's laws could be relied on-whether they were universal for all plants. Were there practical applications: Could the yield and quality of staple crops like maize, cotton, and tobacco be increased by finding the gene, or genes, responsible for these traits and breeding them into the plant? What part, if any, was played by the environment? Could the physical elements of temperature, moisture, and light affect the way genes behave? Even Mendelists conceded that the individual development of an organism must be part nature and part nurture.

When Vavilov arrived at the Petrovka in 1906, Russian biologists, like those in other industrial nations, were split into Lamarckian and Mendelian camps. Some of the older professors at the Petrovka tended to scoff at the new theories of genetics and genetics did not exist in Russia as a discipline; there were no specialized genetics institutions or periodicals.[10] These older academics considered plant breeding to be an ancient art, a native skill born of observation of nature in the raw, not a scientific discipline based on a complex mathematical theory or ratios of dominant and recessive factors. Farmers had been selecting plants for thousands of years and were ideally suited to the task, in their view. Picking the best plants had always been the job of uneducated peasants, not learned academics, and some of the older professors thought that was how things should continue.[11]

In Russia, the Mendel–Lamarck dispute gave rise to an important linguistic distinction. In the past, traditional plant breeders had been known by the Russian word *sortovody*, literally, strain breeders. Now a new term, *selektsioner* (from the Latin *selectio* and meaning, literally, selectionists), was adopted by the new generation of scientifically trained plant breeders. Nikolai Ivanovich was definitely a *selektsioner*. He learned of the importance of genetics for crop breeding from his other progressive teacher, Dionisius Rudzinsky, who in 1903 had set up Russia's first plant breeding station at Kharkov in Ukraine.

The two types of plant breeder often clashed in heated debates at the Petrovka. One summer, Nikolai Ivanovich and several other students were in a private railroad coach traveling from Moscow to Kharkov to attend the First Congress of Selectionists—the full title, reflecting the importance its organizers attached to it, was All-Russian Congress of Activists of Selection of Agricultural Plants, Seed-breeding and Distribution of Material.[12] The students in the coach got into a furious argument about Mendelian laws and their relevance to plant breeding, so heated that it threatened to become violent. Nikolai Ivanovich stepped in, suggesting that they should put Mendelism on trial. He organized a mock court hearing in which he appeared as Mendel's defender, promoting plant selection as a "science" rather than an "art." Witnesses were called for both sides. The prosecution opened the "case" by declaring that farmers had been selecting their best crop plants for thousands of years, challenging Nikolai Ivanovich to "prove" how the young agronomists were in a position, at the beginning of the twentieth century, to know better than the peasants with a good eye for a good, healthy plant, or a fine cow.

Vavilov argued with great passion that in 95 percent of the cases, the peasant farmer did not improve the yield of his crops, or the milk production of his cow, because he was not aware of Mendel's laws governing the inheritance of characteristics. He had no idea which of the traits he had selected would continue on in future generations and which would simply disappear. Applying Mendel's theory of dominant and recessive genes, a botanist could now forecast which characteristics of his crop plant would continue, and which would not. The jury "voted" unanimously in favor of recognizing plant selection as a science and forecast a long and successful career in plant selection for Nikolai Ivanovich.

In private, however, Nikolai was not so sure. To be a good *selekt-sioner* required, he confided to his diary, "more brains and talents of all sorts" than he felt he could muster.[13] "In addition to having a sharp eye, it was necessary to master languages to read foreign publications, to have acquaintance with mathematics, to possess patience, endurance and a youthful thrust." He doubted that he could put all these virtues together.

Even so, he would become an expert observer of the variations in plants and which variations might signify something useful—despite the slightly blurred vision in one eye from the home chemistry experiment. The more specimens he collected, the more fascinated he became with the sheer scale and complexity of the problem. Most of us looking at a field of ripe wheat, for example, see only waves of amber grain, as the anthem goes. But plant breeders with a keen eye, like Nikolai Ivanovich, see many variations that distinguish one stem of wheat from another and give an indication of the plant's worth. The height of the main stem itself can range from roughly one to four feet, and this becomes important when considering whether the stem is capable of bearing lots of grain without falling over. The stem can have anything between seven and nine slender green leaves, and the leaf angle can be almost vertical to the stem or at various points until it is horizontal to the stem, or even pointing downward. This is important when a plant breeder is considering how many plants would fit into a given space. A closer inspection of the wheat stem, perhaps with a magnifying glass, may reveal still more distinctions. For example, some have little hairs at the junction of the leaf and the stem, and some do not. Most ears of wheat have what look like wispy beards, or awns, growing out of the shell of the seeds. A few wheat plants are awnless, with no beard, and the presence of awns appears to increase yield potential in relatively warm climates. Plant breeders are always looking for signs of disease, little yellow spots of powdery mildew, or brown and red spots, known as rust, on the leaves. If plants reach maturity without mildew or rust, it's a sign they might have a genetic makeup resistant to such diseases.

In his final year at the Petrovka, Nikolai Ivanovich worked on the Poltava Experimental Station in Ukraine where he conducted experiments on disease resistance in oats, wheat, and barley, and at the Petrovka's experimental station on plant immunity. He graduated in the spring of 1911 with a first-class degree in agronomy, although his thesis reflected his childhood passion for zoology: the frogs and other slimy creatures of the ponds of Middle Presnya. His thesis, and first published paper, was entitled "Field Slugs as Pests to Winter Crops in Moscow *Gubernya.*" He was disappointed with himself for failing animal husbandry—but his excitement about the future soon overcame his gloom. "It's a trifle, after all," he wrote in his diary; "tomorrow everything will appear in a rosy tint."[14]

According to the custom for Russia's star graduates who were going on to become academics, Vavilov would now include study in Europe, either in German, French, or British laboratories, but he would not go alone.

The Petrovka's female students were especially attracted to the handsome, skinny, almost diffident youth in his elegant suits as he dashed from class to library, from scientific debates to extra language lessons.

One female student in particular caught his eye. Her name was Yekaterina Nikolayevna Sakharova. The daughter of an accountant and from a higher social stratum than Nikolai, she was one year ahead of him at the Petrovka. Katya, as she was known, was rather plain and very serious, even stern, and it was not her looks, but her impressive erudition in one so young that immediately attracted Nikolai. Katya had graduated from the Fourth Moscow Women's Gymnasium with excellent grades in Russian language and literature, French, German, history, geography, mathematics, and the natural sciences.[15]

She was widely read in the classics of European literature, which she liked to quote, and had also traveled in Germany and the Austrian Tyrol. Her parents had died early; her father in 1904 and her mother four years later. In the aftermath of the 1905 uprising she had come under the influence of her sister, Vera, whose radical politics landed them both in jail. Katya was arrested for belonging to the Social Democrats, the Marxists who eventually split into Bolsheviks and Mensheviks. She was interned for five months before being released to attend the Petrovka.

Unlike Nikolai, whose high school curriculum had concentrated on business studies, Katya arrived at the Petrovka an altogether more rounded person. She was already firmly convinced that she wanted to devote her life to agriculture. Nikolai met her soon after he arrived at the Petrovka and was attracted by her obvious intelligence and the strength of her commitment. For her part, she liked the attention he paid to her and admired the energy and determination he applied to his studies, but the relationship was not romantic. He reached out to her as a child might to his mother or father, neither of whom, in his case, had provided him with the consolation he needed.

The image of himself that Nikolai Ivanovich presented to Katya was much different from his public persona. He did not have any "more precise or clearer aim" than any of his colleagues, he told her.[16] "There are some lights shining vaguely in the mists (forgive the unaccustomed poetic turn) that are luring me on [but] I will not conceal from you that I have very little confidence in myself or in my powers. Occasionally these doubts affect me sharply, more powerfully than it appears to an outsider." Outsiders saw little evidence of physical attraction and were surprised when the couple announced their engagement after she graduated in 1910.

Despite her excellent academic record Katya was not asked to stay on for higher academic work. The tsarist ministry strongly disapproved of the radical politics she espoused and her period of detention could not have helped her career. Instead, she worked as a home tutor outside Moscow.

Nikolai graduated a year later, in 1911, and Pryanishnikov kept him on the staff of the Petrovka preparing him for a professorship. He told Katya that he thought Pryanishnikov had overestimated his talents, especially when he was asked to give the graduation address for the women at the Golytsin Higher School Agricultural Course where he had been a part-time teacher. "I was dumbfounded," he wrote. "My failures in teaching put me in a foul mood and discourage me."[17]

Nikolai Ivanovich had already decided to devote his life to applying science for the public good. He was about to launch himself into a career of plant hunting and breeding and had selected two older mentors, one Russian and one English.

The Russian was Robert Eduardovich Regel, a gentleman horticulturalist of German descent. Regal was an expert in strawberries and the grafting of fruit trees, and a consultant to noblemen owners of large estates. He was the only doctor of gardening in Russia and an active member of the Emperor's Society of Horticulture. The society was founded by his father, Eduard, the director of the St. Petersburg Botanical Gardens, second only to the Royal Botanic Gardens in London in the richness of its seed collection and its herbarium. But Robert Regel was also a practical scientist. After the land reforms of 1906 broke up the old communal farming system and encouraged the more prosperous peasants to become independent farmers, Regel became head of the new Bureau of Applied Botany in St. Petersburg where he started the first collection of Russian cultivated plants.[18] He was also a man of considerable influence in high tsarist circles as a member of the Scientific Committee of the Ministry of Agriculture.

The Englishman was William Bateson, a zoologist and evolutionary biologist and the greatest promoter of Mendel's theories of heredity after Mendel's work had been rediscovered in 1900. Bateson was a great independent thinker who got himself into a lot of trouble with his colleagues by being too dogmatic, especially about his opposition to Darwin's theory that some variations in plants and animals were due to environmental influences. He was convinced that Mendelian genetics would provide a better explanation, and he turned out to be correct. Nikolai Ivanovich set his sights on these influential scientists with a determination and confidence that contrasted with the self-doubt expressed in his letters to Katya.

He had met Regel in Kharkov when he had staged the mock trial of Mendelism on the train. In 1911, he wrote Regel asking for an internship at the Bureau of Applied Botany. The tone of the letter, an appealing mixture of flattery, commitment, and cooperation, displays an extraordinary worldliness for a twenty-four-year-old.

"At the Kharkov Breeding Congress you let me hope for assistance; now dare I repeat again my humble request to let me work with the Bureau . . . so far the only institution in Russia that combines studies in taxonomy with those of the geography of cultivated plants. . . . I would regard any advice from a Bureau staff member regarding the use of your library as really valuable to me. . . . Being clearly aware of the load of work at the Bureau, I would do my best to be of minimal personal burden for the Bureau's employees. I would bring the most necessary instruments (magnifying glass, microscope). I reconcile myself to possible inconveniences in advance."[19]

Regel accepted him. "We expect you in Petersburg in the nearest time. P.S. We have a spare microscope. It would be good if you could bring your magnifying glass with you."[20]

It was the start of a year's research, mostly into wheat. Regel was instantly impressed by Vavilov, but the young agronomist was still unsure of himself. He told Katya that he wanted to be an academic, but he feared that he had too few gifts. "Disillusionment and reversals" were possible.[21] He was especially uncertain about the next phase of his doctorate, an extended trip to European laboratories. "I'm not brave. And I have little confidence that I'll be able to do it.

It's all too fast. It smacks of careerism, which God forbid. All these public appearances are nothing but trouble and dismay." Worst of all, he wrote, he was behind in his scientific reading. "I'm not even dreaming about Mutation Theory. Total ignorance in systematics and inability to experiment at all. And my language is horrible. I have to study and study."[22]

It was as though the two youngsters had already set the boundary of their relationship; that it would be about friendship, moral support, and, above all, science, not love. Katya helped him with private moments of self-doubt and Nikolai would be able to supply her with something she did not have—a family, and a home. They were married in 1912 in Moscow. The picture taken for their wedding shows them both looking intently into the camera; Nikolai looks apprehensive and vulnerable, Katya looks almost uninterested, as though she were going through some enforced performance not to her liking. Neither of them wanted any fuss made of their marriage and Nikolai contrived to keep it a secret from his colleagues at the Petrovka. But one of the professors found out and "turned him in," as a student would later recall, so that they could celebrate.[23] Shortly afterward the couple set off for Europe, to Nikolai Ivanovich's second chosen mentor, William Bateson. It was a trip that would be the making of Nikolai as a scientist, but would increasingly isolate Katya from her scientific work, and from him.

CHAPTER 3

In Darwin's Library

With his grand theory of evolution, Darwin had not explained the mechanism of heredity or the source of the multitude of variations found in living organisms. The question for geneticists was how to meld Mendel's theory of heredity with the fact of evolution. At the turn of the century, William Bateson, the fiercely independent English evolutionary biologist, was the preeminent promoter of Mendel's work and became a firm friend of the young Vavilov.

Nikolai Ivanovich could hardly contain his delight when he arrived at the Botanical Institute in Cambridge and was allowed into the inner sanctum of world botany—Darwin's personal library. As he opened the rare volumes of nineteenth-century European plant hunters, he saw pages annotated by Darwin, and Nikolai felt the great investigator himself at work. It was possible, he would write, to trace Darwin's "creative path, the vast painstaking work that preceded his theories of the evolution of species."[1]

He marveled at the extent of Darwin's research, covering every cultivated plant—"maize, potato, green vegetables, fruit, berries . . .

carefully assembling facts on the thousands of varieties of roses, the evolution of the gooseberry from a wild one weighing half a gram to one weighing 53 grams, or the amazing range of variation in the pumpkin where a cultivated variety was a thousand times larger than the wild one." In his appreciation for being allowed to see such treasures, Nikolai noted, "Darwin used to joke that he was a millionaire of facts."[2]

Beginning in 1913, the young Russian agronomist traveled across Europe to learn the state of science outside his homeland. In a frantic twenty months, he swept through the libraries and laboratories of some of the most progressive biologists in England, France, and Germany. At each stop, he presented his calling card—an introduction from his Russian mentor, Robert Regel, whose name still registered in these important circles as the premier plant breeder of Russia. Regel's introductions were priceless because they allowed Vavilov to meet not only the great theorists of the new science of genetics, but also the practitioners, the ones who were turning the theories into improved plant varieties.

Discussions with the great genetic researchers of the day brought Nikolai Ivanovich up to date with current thinking on the problem of heredity and variation, the central problem of all biology. The young Vavilov absorbed this knowledge quickly and eagerly, and despite his youth—he was a generation younger than his interlocutors—he was treated as an equal. As he moved from one famous laboratory to another, the young Russian was excited to meet these icons of European biology, and, at the same time, impatient to return home and start bringing Russian biology up to the European standard.

In Britain, he worked with William Bateson, a key figure in the rediscovery of Mendel's work in 1900, and Mendel's most fervent supporter in Britain. A classics scholar, Bateson coined the term "genetics" from the Greek, *genno*, "to give birth," and thus defined the new science. In Germany, Vavilov visited Ernst Haeckel, then eighty years old and the first scientist to declare that the nucleus of the cell includes material (he did not call them genes) that acts as the basis of heredity. Yet, Haeckel also believed that the cytoplasm, the material that surrounds the cell, played a key role in heredity by transmitting characteristics to the next generation that were acquired during the organism's lifetime—the theory of the French biologist Lamarck.

Regel's contacts also introduced Vavilov to scientists who were using the new theories to breed better plants. At Cambridge, Nikolai Ivanovich worked under Rowland Biffen, who was the first to apply Mendel's work to crop development. Biffen was key for plant breeders because he showed for the first time that resistance to disease was inherited according to Mendel's laws. Biffen also said that the process was controlled by "factors of inheritance" carried by the chromosomes. [3]

In Paris, the young Russian worked as an intern at the Pasteur Institute. For several more weeks, he worked near Paris with the agricultural firm of Vilmorin and Andrier, France's most important seed merchants. (At Vilmorin's, the handsome young Vavilov apparently made quite an impression on the owners and especially on the owner's wife, Madame de Vilmorin. Her connection with the French government would later be of great use to him in obtaining visas for his plant hunting expeditions.)

For all this frenzied movement from one laboratory to another, Vavilov's most intensive and productive sessions were undoubtedly with Bateson. Crusty and controversial, Bateson founded what became known as the Cambridge school of genetics, thrusting the genetics debate into the university's common rooms and dining halls. In 1910, Bateson became the director of the premier European center of genetics, the John Innes Horticultural Institution at Merton, outside London. He moved the debate about plant genetics out of the greenhouses to practical applications: how agriculture could be adapted to feed the world's hungry—Vavilov's territory.

Despite their age difference—Bateson was twenty-six years older than Nikolai Ivanovich—the two scientists quickly became friends, and Bateson was to have a tremendous influence on Nikolai's career, not only on his attitude to science and his research projects, but also on the administration of science, the way Nikolai would eventually run his own Institute.

At Merton, Bateson ran a group of researchers devoted to Mendelian genetics. He abhorred Lamarckism and the inheritance of acquired characteristics, but he was also a skeptic of Darwin's natural selection theory of evolution. Like other biologists he had difficulty accepting the concept that small variations could lead to big evolutionary changes and new species. Was it not more likely that some other method, a massive single change or mutation as it came to be called, might take place?

Bateson, a zoologist by training, had noticed what he called "discontinuous steps"—big changes in variations as opposed to the small continuous steps put forward by Darwin. Darwin had also noticed these discontinuous steps, but had assumed that they would be discarded by creatures and plants through "natural selection." Bateson thought that the "discontinuous steps" were probably how evolution happened—in fits and starts, he suggested, not gradually or continuously.

Bateson also had problems with Darwin's idea that organisms were in a permanent state of competition with one another, that there was a continuing struggle that was resolved by the "survival of the fittest." Another of Bateson's cherished examples was in his own garden where he grew three species of veronica, a perennial garden plant with spikes of flowers, usually in blues, purples, and pinks. Bateson grew the plants side by side and would tell anyone who would listen that each of them had distinguishing features—or variations—that did not disappear because of some survival mechanism, but continued generation after generation.[4]

We can imagine him forcefully lecturing his young Russian visitor, "It is clear to me, Vavilov, that the variations of the veronicas have not arisen on account of their survival value, but rather because none of their variations was so damaging as to lead to the extermination of its possessor. Tolerance, my dear Vavilov, has as much to do with the diversity of species as selection." In other words, veronicas had several varieties that lived happily forever, they tolerated their surroundings, and did not evolve. Their genetic makeup did not change.

During these months at Merton, the younger Russian and the older Englishman sat for hours discussing evolution, heredity, and variation. They could be found on many evenings in Bateson's study, with Bateson in his tweed suit and bow tie, puffing on his cigar, and Nikolai Ivanovich, in his formal gray serge suit, white shirt, collar, and tie. Bateson's wife, Beatrice, might be playing her cello in the next room, as the scientists carried on their discussion into the small hours, about how these variations might be linked, about the influences of the gene, and the part played by the environment, or even by the nurturing farmer.

Bateson relished the role of the skeptic, always ready to challenge any theory that landed on his desk. Genetics was certainly not a perfect solution, he admitted. If one wanted to construct a true synthesis of evolutionary theory, the professor often said, one that combined Darwin and Mendel and all knowledge of the inheritance and development of an organism, it was necessary that variation and heredity "be minutely examined." It was just the kind of encouragement that Nikolai Ivanovich sought. Minute examination of variations in cultivated plants was exactly what he yearned to do, what he was determined to make his life's work when he returned to Russia.

Bateson also encouraged Vavilov to keep an open mind to possibilities that appear to be counterintuitive. Occasionally, and mischievously, Bateson turned the established idea of evolution upside down, just to emphasize his point that none of the current theories was absolute, that all was not yet decided by any means.

When the facts of genetics were more widely debated, Bateson undoubtedly suggested, there would be many thoughtful discussions on this matter. As Bateson had told others, "I ask you simply to open your mind to this possibility. It involves a certain effort."[5]

Such openness, such deviation from the norm, was so attractive to Nikolai Ivanovich that he was reluctant to leave after a year at the John Innes institute. For him, Bateson and his methods were a revelation. Russia's institutions were rigid and hierarchical with little real freedom for a scientist who suddenly found a new opening to leap quickly from an assigned task into researching something new, or merely promising. At Merton, there were all sorts of research projects at all stages of experimentation—from wheat, flax, gooseberries, primroses, begonias, tobacco, and potatoes, to snapdragons, plums, apples, and strawberries, and to rabbits, chickens, canaries, grasshoppers, and cockroaches.

Bateson gave his researchers the same kind of freedom he demanded for himself. The professor bridled at the tyranny of orthodoxy and was always seeking alternatives, those unmapped detours that sometimes brought the best scientific rewards.

"Contrary to a traditional concept of the reticent English personality, it was impossible to imagine a more cordial attitude, attention and readiness to help, encountered by a young Russian researcher in Merton," Vavilov wrote later.[6]

Professor Bateson and his researchers indeed gave Nikolai Ivanovich a headful of ideas to take home. The professor encouraged Nikolai to launch his plant hunting expeditions to look for new varieties of crop plants, and the professor's research agenda at the Merton institute would inspire him to set up a similar institution with a broad range of research topics, autonomy of the researchers, and a director with a strong personality.

Meanwhile, Katya was not inclined to use their travels to further her own agronomic expertise. She preferred to study English literature. While Nikolai studied the immunity of wheat to rust and mold, she buried herself in the popular literary works of the day—H. G. Wells, Thomas Hardy, Rudyard Kipling, George Bernard Shaw, and James Joyce as well as American authors such as Theodore Dreiser and Ralph Waldo Emerson. She also read the German philospher Oswald Spengler. Katya had an amazing memory and, rather irritatingly for her companions, a habit of quoting passages from her literary discoveries in everyday conversation. This oddity only served to drive people away, including her husband. As she became more and more isolated from his work, the thing that had brought them together in the first place, she had nothing to give him as a substitute and was thus unable to supply him with the intellectual consolation that he sought. Nikolai Ivanovich retreated even more into his science.[7]

The young couple's stay in Europe was cut short by the start of World War I in August 1914. Gathering his precious samples of rust-resistant wheat from the Cambridge and Merton labs, he and Katya rushed home by train. Their heavier bags, including research books and botanical specimens, went by sea and were lost when the ship carrying them was sunk by a German mine. It was a terrible loss, but Nikolai Ivanovich had established such a great rapport with his new mentors like Bateson and the Vilmorins that he was able to replace most of his bounty.

In Moscow, all men of military age were now subject to the draft—millions were being called up. Nikolai's brother, Sergei, was among the first to go. Fortunately for Nikolai, his childhood chemistry disaster that resulted in blurred vision in one eye spared him from active service. Instead, he threw himself into the next stage of his academic career, his doctoral degree and research on disease-resistant crops at the Petrovka's experimental farm. After months of work, he finally found a wheat variety that wouldn't die even after being doused with mildew. It was his first intoxicating plant breeding success, discovering a food crop that could survive a fungus. And it would encourage the young specialist to find other plants that might resist other diseases. For that, the young Vavilov would have to follow in Professor Bateson's footsteps. He would have to go to Turkmenistan and Kazakhstan and the Caspian and other fertile areas to search for plants containing the magic genes. There he would search for wheat or rice plants that might survive in cold or heat, too much water and too little, as he began his adventures as a world-renowned plant hunter, looking for the secrets to overcome famine.

CHAPTER 4

Moscow, Summer 1916

In the Great War of 1914–18, the tsar would mobilize fifteen million men, mostly peasants. Across the vast lands of the Russian empire, conscripts were transported thousands of miles to the eastern front, or sent south into Persia to meet Turkish forces. The war affected the Vavilov brothers in very different ways. One would be sent to the eastern front, the other to Persia.

Nikolai's mother, Alexandra Mikhailovna, was in the kitchen preparing breakfast at No. 13 Middle Presnya one June morning, when the front door burst open and Nikolai Ivanovich, a startling vision in a brand-new cream linen suit, white collar, and tie, bounded into the house. He announced that a car from the Ministry of Agriculture was coming to pick him up after breakfast. Then he was gone.[1]

Alexandra Mikhailovna must have muttered to herself that soldiers normally go to war in battle dress, not cream linen suits, and on foot or horseback, not in chauffeur-driven cars. When Sergei, her younger son, had departed for the eastern front two years earlier, she reminded herself, he had worn the field gray uniform, epaulets, and high boots of the tsarist officer class. Poor boy, he was not cut

43

out to be a soldier, she had fretted. During the army physical he had complained about being made to strip naked and inspected "like a horse." "I am only slightly like a 'horse'—my muscles are weak and it would be useful to become slightly more 'horsey,'" he had written in his diary.[2] He was not a fighting man, not like Nikolai Ivanovich, who was always ready with his fists. Sergei had studied physics, but for the army he was relegated to a signals regiment where he specialized in little more than the Morse code. He longed for home. "First of all I got mixed with 'the people,'" he wrote in his diary, "all the rank-and-file mob with whom I sit in the shack now, likeable people, often rude. Everything in life is crippled. Horror turns into constant sickness. Filth, boredom of war. . . . I am not a soldier, neither in my body nor in my soul."[3]

Alexandra Mikhailovna had read and reread his letters so many times she could quote from them. She had no idea where he was on the eastern front; he was not allowed to write about his duties. He could only tell her, "I decipher and then I cipher."[4] His mother dutifully sent food parcels and magazines about physics to her son, who wrote about longing to be in Presnya "in a quiet room where the sun will rise again over my personal life."

Nikolai Ivanovich, who had spent the early war years living with his wife, Katya, next door to his parents' house, was finally being mobilized, but instead of being called into the regular army, he was merely being consulted because of his expertise as a botanist. The soldiers at the Persian front had been getting sick eating the local bread. It made them dizzy, unable to focus. Alexandra Mikhailovna knew Nikolai Ivanovich had been overjoyed about being asked to solve this mystery and could barely wait to leave for his part in the war effort. She was certain that her son would make a valuable contribution, and she was sure he would be able to look after himself. He was so much more resourceful than his brother, stronger in every way than the gentle Sergei.

Even so, she was overcome with the sadness every mother comes to know when a child leaves home. Nikolai Ivanovich was twenty-eight and had always come back after the field trips to the east and even the exciting year in Europe. Now she feared that he might be leaving, perhaps even for good. She knew about his plans to organize plant hunting expeditions that would keep him away for months, maybe even years. There would be nothing to hold him in Presnya now, not his parents, not his work, and certainly not his marriage. There was no love that she could see between Nikolai and Katya, no warmth, not even the moral support that was so important to him in the beginning.

Within their family circle, Katya had become known as a peculiar and difficult person. She was clever and she knew several languages and Nikolai undoubtedly valued her intellectual talents. But Katya had made it clear that she considered herself above the Vavilovs and their friends, too smart for people barely one generation away from the fields. She was condescending, priggish, and aloof, and her favorite phrases were dismissive: "I have no idea." "It's not my business." "I couldn't care less."[5] She would lecture Nikolai Ivanovich, like a professor, not a wife, observers would later note. Even Sergei expressed surprise that Katya had failed to recognize that his older brother was a genius. Alexandra Mikhailovna had often wondered why Nikolai Ivanovich had stayed with her for four years. Now she feared he might use his call-up as an opportunity to leave his wife and his family behind in Presnya.

Alexandra Mikhailovna would have faced this fear with prayers, like many of her class at this frighteningly chaotic time in Russia's history. She believed in God, an uncomplicated faith involving regular observance of the fasts and festivals of the Russian Orthodox Church, and beginning and ending each day with a prayer.[6] Her husband, Ivan Ilyich, was religious in the same way; he followed her routine, but it was hard for her to determine whether God and the

Church were more significant in his life than in hers, as Ivan Ilych
found it difficult to talk about such things. She had to keep her wor-
ries to herself and ask God to protect her sons at war. Perhaps that
day she offered up her prayers in the garden, a cheerful place that
June with new apple blossoms and fresh grasses underfoot.

The mother of these two budding scientists did not fully com-
prehend the tumult in Russia. Presnya was the only place she had
known, since she was not able to travel like her husband and sons
and had no professional status, like her daughters, who were medical doctors. Since the failed uprising of 1905 had been fought outside her front door, talk of war and revolution had been constantly
in the air. And each day the news from the front lines grew worse.
By the summer of 1916, the second full summer of the war, the
Russian Imperial Army had suffered catastrophic casualties, with
nearly a million dead, wounded, or missing and three quarters of a
million taken prisoner. The tsar had mobilized another four million
men, mostly peasants, adding to the eleven million already drafted.
Across the vast lands of the Russian empire, conscripts were transported thousands of miles to the eastern front, or sent south into
Persia to press attacks on Turkey.

Since coming back from Europe in 1914 and working on his
doctorate at the Petrovka, Nikolai Ivanovich would always man-
age to rise above these upheavals by single-mindedly launching
himself into his science. While rebellions in Russia were brewing,
he won awards for his work and by 1916 would be known as one
of Russia's most promising young scientists. He roamed across the
Russian empire looking for rare varieties of wheat and soon had
created a collection that was the envy of wheat breeders in Europe
and America.

Robert Regel would write to the Ministry of Agriculture, "In the
past twenty years the problems of plant immunity have been studied a great deal . . . [but] I dare say no one has ever approached the
solution of these complex problems with such a breadth of views

and comprehensive coverage as Vavilov has done . . . his work is an honor to Russian science."[7]

When Nikolai Ivanovich had received the Persian assignment from the ministry, he had assured his parents that he did not expect to be away for long—a few weeks at the most. He said that he was already certain he knew what the problem was and would be able to fix it in short order.

As she prayed for the future that day, Alexandra Mikhailovna would certainly have added a plea for the present, for that week. She worried like everyone else in Moscow in those days that there would be enough food for the extended family. The loss of Poland, the western Ukraine, and parts of the Baltic provinces to the Germans—all good grain-producing areas—had already had a severe impact on Russia's harvest. The millions of peasants taken from their work in the fields only made the situation worse.

The railway network could not cope with the increased traffic due to the war, and soon all commodities were in short supply—fuel, textiles, and even Ivan Ilyich's precious tobacco. Prices had been rising in the cities, and the rural provinces around St. Petersburg (renamed Petrograd in 1914) and Moscow barely had food for those left behind.

That morning, after breakfast, the car arrived from the ministry. Ivan Ilyich stood, ramrod-straight as usual, at the front door of his house. He wore his customary fine gray woolen suit and a matching fedora.

Members of the extended Vavilov family had also come to bid farewell—the lodger Rubtsova, a young woman doctor, Marina Pavlovna, and the family piano teacher, Dubinin, nicknamed "Lion's Mane" for his lavishly unkempt hair. He taught only the girls; Ivan Ilyich said a music education was not appropriate for boys.[8]

Next door, outside No. 11 Middle Presnya on the corner of Predtechensky Lane, a one-story wooden house with a mezzanine, even Katya emerged to say goodbye.

Nikolai Ivanovich was walking between the two houses, in his new suit and brandishing his topee, a khaki pith helmet of the kind worn by British officers in Africa and India.

"This is my hello-goodbye hat," he announced playfully to his nephew, Alexander Ipatiev.[9] As he had done on many occasions, Nikolai Ivanovich chased his nephew around the garden as they recited Alexander's favorite proverb:

> *The bird catches the worm,*
> *The cat has the bird for breakfast,*
> *The dog gobbles up the cat,*
> *A hungry wolf devours the dog,*
> *But what's the use of it?*
> *The mighty lion ate the wolf,*
> *And a man, spotting the lion, shot him,*
> *The worms ate the lion.*

They had recited the proverb many times, and to Alexander Ipatiev, this was another treat from Uncle Kolya, as he knew Nikolai. This uncle, he would later say, was the idol of his life, a man he looked up to more than any other.

Nikolai supervised the driver packing his trunk into the car, slung his brown leather plant specimen bag over his shoulder, hugged each family member in turn, and then he was gone. From Middle Presnya, the car headed across the Moskva River to the Kursk railway station and the train that would take him to Ashkhabad in Turkmenistan. There, several days later and almost two thousand miles from Moscow, he would buy horses and travel to the Persian border. Nikolai sat back in the seat and smiled as he began his tour of duty. He was not going to the front to shoot Germans, Austrians, or Turks; he was going to collect plants, a larger mission than fighting one world war.

Far from being frightened at the prospect of the front line, this was the opportunity he had been dreaming about for some time. It was an officially financed trip into northern Persia where, if his calculations were right, he expected to find a rare type of wheat with extremely high resistance to the powdery mildew that so devastated crops in northern Russia.

The poisoning of the troops was not such a big mystery, he guessed, and he probably could have diagnosed the problem correctly from the Petrovka. During his study of crops in the Cis-Caspian, on Persia's northern border with Russia a few years earlier, the young Vavilov had found wheat infested with a poisonous Darnel ryegrass, *Lolium temulentum L*, that could produce the drunkenness symptoms similar to the ones reported by the army. Vavilov quickly solved the "mystery."

Although his mission was military he insisted on retaining his civilian status, carrying orders from the Ministry of Agriculture, not the Ministry of War.

In Persia, Vavilov bought three horses, hired a guide, and headed into the country's interior in the direction of Hamadan and Kermanshah. The summer air was filled with the pleasant scent of *shabdar*, the Persian clover used as forage. In the mountains around Menzil, he came across thickets of wild perennial flax. In Gilyan province, he found rice of an especially high quality, grown in fields remarkably free of weeds. Nikolai Ivanovich eagerly scooped up samples of these new seeds to take back to Moscow.

His peaceful pursuit of plant hunting was constantly interrupted by the war going on around him. A detachment of Cossacks, patrolling the front lines, arrested him and accused him of being a spy for Germany. Although he was carrying official papers from Petrograd, the Cossacks were suspicious. It did not help that Nikolai Ivanovich had written up his field notebook in English and that several of his reference books were in German. A Cossack officer who announced that his assignment was to "destroy vermin" interrogated Nikolai Ivanovich. The officer seemed much more interested in proving that he was a spy than finding out if he was a bona fide researcher and there was a good reason. The Imperial Army was paying a bounty of up to a thousand gold rubles for apprehending German and Turkish agents. The Cossacks held him for three days while they checked his credentials by telegraph with Petrograd. When confirmation of his official status finally came from the Ministry of Agriculture, the Cossacks released him. He was free to go plant hunting.

CHAPTER 5

On the Roof of the World, 1916

At the beginning of the twentieth century, plant breeders began to search for rare genes in the wild ancestors of the major crop plants. These genes might give staples like wheat and barley resistance against disease and extremes of temperature and drought. Nikolai Ivanovich had a theory that the best place to look was in remote high mountain ranges, such as the Pamirs, on the southern border of the Russian empire. Instead of returning to Moscow after Persia, he bought three pack horses and launched his first foreign expedition.

Winter came early to the Pamirs in September 1916. Snow flurries covered the mountainous steppes and narrow gorges of central Asia where China, India, Russia, and Afghanistan meet. The Demri-Shaurg glacier was covered with a treacherous moraine of slippery gray shale, requiring those on horseback to dismount and proceed on foot.[1] Even a path around the edge of the glacier had its dangers. Ice bridges over rivers and streams had become unstable in the summer melt, and now gave way under the lightest foot. To cross the swift but icy waters, a traveler had to build rafts, or *gupsars*, of tree trunks

51

buoyed by inflated goat skins. Where the route's rocky ledges were too narrow for a foothold, there were still the *ovrings*, made from wooden poles driven into the rock and covered with branches and flat stones. These pathways, created by the tsar's troops, afforded precarious support at best, especially for a pack horse. Over this dangerous trail, it was not possible to cover more than a mile in an hour, and many travelers never arrived at their destination. For good reason, the Pamirs were known as the foothills of death.

In this third year of the Great War there was a new danger. The front line between Russian and Turkish forces was in northeastern Iran, several hundred miles to the west, but a recent attack by the Turks had forced the Russian army to withdraw. Shocked by the defeat and needing local reinforcements, the tsar had ordered mobilization of the nomadic Kirghiz tribesmen, but they had refused to join the Russian army. The tsar sent his fiercely loyal Cossacks to round up the tribesmen, but they had fled into the Pamirs. There had been reports of travelers being harassed, robbed, and even murdered. But Nikolai Ivanovich was determined to press on.

His studies of ancient agricultural patterns had suggested to him that the high fertile valleys of the Pamirs contained hardy varieties of wheat as yet unknown to present-day botanists. He was developing his theory about the origins of cultivated plants and he believed that other plant hunters from Europe and America had been looking in the wrong places for unknown varieties. They had concentrated their search on the valleys where the early civilizations began—the lowland river plains of the Tigris and Euphrates. But Nikolai Ivanovich imagined that the greatest variety could be found in mountainous areas such as the Pamirs. He argued that the early farmers, in their struggle for land and existence within densely populated areas, had been forced to settle at almost inaccessible altitudes. And if he was correct, some genetic treasures would be found in the mountainous areas of southwestern Asia, the mountains of Africa, the Cordilleras of America, the central Asia highlands, or the alpine Caucasus.

The Pamirs straddled the khanate, the princely dominion, of Bukhara (today's Tajikistan), Afghanistan, and China. The stunningly beautiful glacial valleys at between twelve thousand and fourteen thousand feet are surrounded by twenty-thousand-foot-high peaks of the Asia mountain chains radiating out to the east, south, and west.

In the high plateaus, Vavilov expected to find an array of cultivated plants that were adapted to the stony soils, harsh climate, sparse rainfall, and short growing season, almost a replica of the climate of northern Russia.

The obvious routes into the mountains were blocked by the rebellious Kirghiz, which meant that the only way was the most difficult—over the Demri-Shaurg glacier. At the imperial military headquarters in Bukhara, the local Russian governor and the military commander warned sternly against such folly. The early snows had already made the trip hazardous, there were few villages, and the maps were inadequate.

Nikolai Ivanovich, facing a choice between returning to Moscow and taking the route across the glacier, insisted on continuing his mission. He requested a military escort, but the officials scoffed: he was not on official government business this time and only had letters from the Petrovka and the obscure Moscow Society of Investigators of Nature. The military commander refused to spare a single Cossack.

The Russian political agent in Bukhara, hoping that Nikolai Ivanovich would mention him kindly in dispatches that would reach the government in Petrograd, came to the rescue. He arranged for the emir of Bukhara to send word to his underlings along the glacier route that if anything bad befell the Russian scientist they would be hanged. He also offered to arrange horses and a guide, Khan Kil'dy Mirza-Bashi. A man of about fifty and overweight, he wore an oriental robe embroidered with colossal multicolored flowers and a shiny silver belt. Nikolai Ivanovich later confessed to being somewhat intimidated by the splendid attire of his guide. "It appeared that it was not he who escorted me, but I him," he wrote.[2] And he worried that this fat person would have trouble negotiating the tricky mountain passes.

The choice of Mirza-Bashi turned out to be an excellent one. He was a personable, educated man—indeed his name translated as "The Learned One." He spoke the Uzbek version of Farsi, the native language of Persia, understood the Kirghizian dialect, and even knew some Russian. Although ignorant of botanical matters, he was eager to learn. He quickly became interested in collecting plants and how to learn from local people about their agriculture. "In general he was not a bad assistant," wrote Nikolai Ivanovich, ". . . although constantly complaining that in all his life, traveling over the mountains of Bukhara, he had never seen a more wretched area."[3]

In the middle of September, Nikolai Ivanovich's little caravan, including two bearers and six horses, set out with only the rough— and often inaccurate—Russian military maps as their reference. A more peaceful spectacle in the midst of war and tribal uprising is hard to imagine. Mirza-Bashi in his flowing robe; Nikolai Ivanovich in a gray woolen overcoat, a three-piece suit, white shirt and tie, and fedora. His leather specimen satchel was placed firmly across his chest and his camera dangled from his shoulder. The pack horses were loaded with supplies and reference books, and their plan was to stop overnight in the tiny villages in accommodations arranged by Mirza-Bashi.

The Pamirs rise steeply out of the Bukharan plain. The caravan followed the six-foot-wide trail cut by the sappers into a perpendicular cliff, several thousand feet high. The horses had to cross icy rivers and chasms, many too wide to jump over. At one such chasm, Vavilov would report, the two guides lay across a three-foot-wide gap to form a bridge while the horses, Vavilov, and the portly Mirza-Bashi walked over them.

The passage over the Demri-Shaurg glacier was slow, sometimes requiring three or four hours to advance as many miles, and the party was forced to camp overnight among the rocks. Nikolai Ivanovich wrote, "We had not calculated on a night camp along the glacier. The lack of warm clothes forced us to start moving early. Almost freezing to death for two days was not very pleasant. We managed to keep going only by a common lowering of expectations; by indifference to all that happened." Their goal became simply survival.

The group finally completed passage over the glacier on September 18, but the next stage was almost as daunting. The trail continued along a steep mountain face some three thousand feet above the valley of the River Pyandzh, the upper reaches of one of the great rivers of central Asia.

In parts of the trail, the party could proceed only on foot, pulling the horses behind them. Sometimes the trail narrowed to single file, sometimes it was wide enough for two or three abreast; but often it was a rocky staircase over which the horses, though accustomed to mountain trails, could move only with great caution. Above them towered the high cliffs and three thousand feet below the icy blue waters of the river. This journey would have tested the physical endurance of the toughest of the tsar's mountain troops, and yet here was Nikolai Ivanovich, in his suit, without any special equipment, leading his group along these paths with the same single-mindedness that he used in any new episode of his life. As they climbed, the path got worse and the precipitous drop into the river valley more vivid. Any sudden movement that would frighten the horses had to be avoided.

At one point, the path broadened and Nikolai ordered the group to remount. But as they turned a corner, two eagles flew out from a nest and circled the party. Nikolai Ivanovich's horse shied and bolted along the trail. He lost the reins. For a few terrifying seconds

Russia's intrepid plant hunter thought that he and his horse would surely fall into the ravine. Somehow the horse managed to stay on the trail and somehow Vavilov managed to stay in the saddle. "Such moments," he noted prosaically, "steel one for the rest of one's life; they prepare a scientist for all difficulties, all adversities and everything unexpected. In this respect my first expedition was especially useful."

The party was not always so lucky. While fording the rapids of a mountain river, one of the pack horses slipped and fell. The churning waters carried the horse downstream—with its traveling trunk of journals and plant specimens. The horse and the trunk disappeared beneath the ice bridge. A search lasting several hours along the riverbed failed to find the horse, the equipment, or any of Nikolai Ivanovich's precious reference books.

After a steep climb the group entered a wide, lush green valley covered with fields and gardens and the regional capital of Garm. The most difficult part of the trail had ended and in some places the horses could even proceed at a trot. Vavilov began collecting samples of local varieties of wheat, barley, rye, and a highly productive crown flax, used in making linen.

In Garm, the caravan rested and reequipped. The two Kirghiz guides insisted on returning home, taking the horses with them, and the mayor of Garm helped to organize a new team of horses with Tajik guides. Between Garm and Khorog, the main military post, the route became easier, a well-beaten path between tiny hamlets of half a dozen houses. Here the Tajik farmers grew spring rye, peas, vetch, and local types of beans and peas. They cultivated the land with simple wooden plows harnessed to a pair of oxen or sometimes even cows.

Nikolai Ivanovich's notes describe the next part of the expedition:

"Soon we were in Shugnan and Rushan with the splendid village of Kalay-Valmar. Wheat that we found at an altitude of about 7,000 feet exceeded our wildest expectations; gigantic rye up to four and a half feet tall and with rigid culms [stalks], large ears and large grains and among it absolutely original forms of so-called non-ligulate rye, undoubtedly initially established there. It turned out that this rye was distinguished by unusually large anthers and large pollen; no doubt an endemic plant! For the sake of it alone, it was worth coming to the Pamirs."

Also in nearby Shugnan, Nikolai Ivanovich met the chief Pamiri plant breeder, Abdul Nazarov, whose wife was an Afghan—from the other side of the Pyandzh River. With his wife's help, Nazarov had managed to obtain seeds of a rare Afghan wheat—which ripened up to twenty days before the Pamiri wheat. It had become the seed of choice for the Tajik farmers, and a few seeds were given to Vavilov for his collection back in Russia.

The breathlessness of these notes gives a clue of his pure scientific excitement, but the energetic Vavilov was, as always, fascinated by his entire surroundings, not only plants but people and language. The people of the Pamirs, he noted, are of Aryan origin with faces that differ little from Europeans. "The personality of the Tajiks is kind and friendly in contrast to Persians, the Tajiks are not timid in front of Europeans; they dress mainly in bright colors. In contrast to the village women of Iran and Darvaz, Tajik women do not hide their faces, although they try to avoid men. The little children are somewhat frightened by a camera."

Nikolai Ivanovich also became an ardent student of what the local people ate. The most common dish was a soup made of peas, barley, wheat, and millet. They ate flat bread, yeast being unknown. They ate meat only on feast days and the ravenous Mirza-Bashi made sure the caravan never missed a feast. On one day, Vavilov wrote that they "traveled ninety instead of the usual forty to fifty kilometers, thus ensuring that we would be in time for it [a feast]." Local chiefs and personalities were always given colorful gowns as gifts and Mirza-Bashi would enjoy not only the food but yet another flowing robe.

A hundred miles farther into the mountains, the party reached Khorog, the most important settlement and the government center of the Pamirs. In the tsarist officers' mess Nikolai Ivanovich was astonished to find "a rather good library" and a piano that had been hauled in by yaks. The officers had made themselves at home, even generating electricity from the fast-flowing River Gunt. "Thus, in the shadows of the Pamirs it was possible to enjoy European conditions for several days," he wrote.

From the ease of Khorog, the expedition struck deeper into the Pamirs along the valley of the rivers Gunt and Shakhdara. And there in all the expanse of this rugged land Vavilov found a prize. It was an endemic wheat with heavy, inflated ears of beautiful white grains that he knew had been caused by the low rainfall. Such a plant might be suitable for the wheat-growing areas of Russia plagued by drought. In addition, the wheat was entirely free of rust and mildew. "No doubt, such wheat had never been seen by any botanist before," he wrote.

Here, in this high plateau, sparsely populated by subsistence farmers, Vavilov had confirmed his initial suspicion about their ancestors; that some of them had fled the plains where wheat and barley, lentils and rye had originally been cultivated by the earliest farmers and had chosen to live in these natural mountain fortresses where they were better protected from wild beasts and attacks from unfriendly neighbors. The Pamirs was not a "center of origin" for these plants, but it was a "natural laboratory" where over millennia "peculiar forms" of food crops had been developed. The great number of wild relatives of these crops also revealed the "enormous plasticity" of the species—the promise of so much more to discover in later expeditions about how these species had evolved. "The presence of wild relatives of barley, wheat, rye and lentils had demonstrated before our eyes that it was possible to solve the most complicated problems of the evolution of plants."

Moreover, Vavilov had seen evidence that a host of different crops—vegetables as well as grain—could be grown at altitudes around twelve thousand feet. Each of these high-altitude varieties appeared to ripen early, grow rapidly, and tolerate the low temperatures experienced during the night, even in summer.

The specimens of cultivated plants that Nikolai Ivanovich had collected, he would write many years later, "far exceeded expectations" and "to a considerable extent determined the direction of future expeditions."

Vavilov's discovery in the Pamirs of new types of wheat and rye had set him on a course of research into the origins of cultivated plants. His first target would be rye, the traditional crop of Russia in pre-Revolutionary times. The traditional Russian black bread is made of rye and, in those times, Russia produced more rye bread than the rest of the world combined. Rye not only produced higher yields than wheat, but it was also more frost-resistant. It was the main winter crop in central, western, and northern Russia, where the climate was severe. He would trace the rye grown on Russian farms from the weed he had found among crops of wheat and barley in southwest Asia. It was the start of what would become an encyclopedic work on the evolution of the world's staples. The question was: where to find what he called these "centers of origin"? And then to test Vavilov's key hypothesis—that the "centers of origin" of these major crops would also contain the greatest diversity of types, and thus a store of invaluable genes.

CHAPTER 6

Revolution and Civil War

The revolution of 1917 was an uprising against the monarchy but also against the appalling losses in the Great War and the chronic shortage of food. In February, the tsar abdicated and a provisional government took over. In October, the Bolsheviks seized power. By March 1918, Russia had signed a peace agreement with Germany at Brest-Litovsk, but at a heavy price, giving up almost a quarter of its territory and its population. There followed a devastating three-year civil war between Bolsheviks, the Red forces, and counterrevolutionaries and foreign invaders, the White forces. Until their defeat in 1920, the White forces controlled much of Siberia and southern Russia. In the midst of this mayhem, Nikolai Ivanovich became a professor of agronomy in Saratov, a grain port on the Volga and a strategic city in the civil war. There, on the edge of Russia's fertile black-earth farmland, he pursued his research, breeding the more productive crop plants that would be desperately needed to offset future famine.

Nikolai Ivanovich returned to Moscow from his travels in the autumn of 1916 amid worsening news from the war front, and mounting strikes and widespread food shortages at home. The end of the monarchy was inevitable; the only question was how. At the Vavilov home in Middle Presnya, Nikolai Ivanovich's father was already planning his exile to Europe, while his mother struggled to find the day's food. Their only source of cheer was that Nikolai's younger brother, Sergei, who had been captured by the Germans near Minsk, had managed to escape from a prisoner of war camp because he spoke such good German.[1] He was coming home.

Each day Nikolai Ivanovich went to work at the Petrovka agricultural academy with renewed enthusiasm, refusing to be distracted by the constant threats around him. He had already decided what his role in life should be. He had a revolution of his own. It was to plant and nurture his tiny green seedlings, mother them to maturity, use them to transform Russian and even world agriculture, and ensure the survival of humanity through an adequate food supply. Life was short; he did not have time, or the inclination, to take part in politics or civil war. He would make use of whichever political structures were in place to further his higher calling. He would use all his ingenuity and resourcefulness to overcome the obstacles, discomforts, and inconveniences put in his way by the tumult of the times. His grand vision demanded concentration and rigorous scientific research.

In the Petrovka's experimental fields, he began breeding the wheat and rye varieties he had brought back from Persia and the Pamirs.[2] By spring, he had published a paper, "On the Origin of Cultivated Rye," which demonstrated that rye had not developed in the same manner as other ancient staples like wheat and barley. Rye probably first appeared as a weed in the mountains of eastern Turkey, Armenia, and northwest Iran, where the harsh climate was unsuitable for wheat and barley, he surmised.[3]

In the summer of 1917, two universities, Saratov and Voronezh in the Volga basin, competed for Vavilov's talents by offering him a full professorship in agronomy, a singular distinction for a thirty-year-old who had only recently obtained his doctorate. The bidding was intense and, in the end, he chose Saratov. Nikolai Ivanovich wrote to a friend, "The fuss is now over, as you can see from the fact that I am living now in Saratov—it's better than Moscow."[4] He was on his own. Katya chose to stay in Moscow—officially because he had no apartment and was sleeping in his office behind a screen. This arrangement lasted for eighteen months; Vavilov was apparently in no rush to change his bachelor lifestyle.

Saratov was founded at the end of the sixteenth century as a frontier post, one of a string of fortresses along the Volga protecting the principality of Muscovy against raiders from the east. In the eighteenth century it was the administrative center of the new Saratov province. In the nineteenth century, Saratov became a boom town, shipping grain and agricultural products. By the early twentieth century it opened a music conservatory, the first provincial art museum with free entry, it had a progressive local government, a broad range of newspapers and publishing houses, and a university, founded in 1909. In the center of town there were impressive churches, several rather grand, neoclassical buildings housing government offices, and a magnificent railroad station. In contrast, along the Volga's right bank, there were strings of workers' wooden tenements.

Before the beginning of the twentieth century, Saratov had already become a center of Russian liberalism with the local intelligentsia sympathizing with the plight of the rural masses. On the eve of the 1917 revolution, Saratov was a center of the Socialist Revolutionary Party, then the country's most popular political party, but after the revolution the Bolsheviks took over and remained in power, despite several internecine struggles and attempts to upend them. Although the White forces never entered Saratov, anyone living in the city was aware of the Bolsehviks' precarious hold. Martial law had to be imposed three times, the first in May 1918 after a mutiny in the local garrison, the second after White forces seized nearby towns, and the third during the White offensive of 1919. During those times, Saratov became an armed camp.[5]

Nikolai Ivanovich arrived in the city a month after the October revolution, and with nowhere to stay he slept in the laboratory of the university's Faculty of Agronomy. The country was veering toward another famine, with farm production disrupted and transportation chaotic. The wealthiest inhabitants were leaving the city and it would become swollen with jobless, hungry refugees from the countryside. In addition to food and housing shortages, there were epidemics of typhus and cholera. Burglaries were commonplace and people who appeared to be well off, and that included Nikolai Ivanovich in his three-piece suits, were in danger of being robbed or molested by drunken soldiers.

The bitter winter of 1918–19 was a time when "each person withdrew into his personal life and concern for a piece of bread," a local newspaper editor wrote. Yet when pressed to take sides, "everyone understood that there could be no waverers; you were either for or against the revolution."[6] Those against were arrested and faced summary execution. Among the intelligentsia, some were cunning enough to be spared.[7]

At first, the university was an island of relative calm. Although the faculty was generally opposed to the Bolsheviks and lectures contained anti-Bolshevik propaganda, the new rule brought few changes.[8] Students who backed the Bolsheviks were censored, and even by 1919 only 4 percent of the ten thousand students enrolled belonged to the communist student union.

The Faculty of Agronomy was already one of the largest in Russia, and Vavilov immediately started a rapid expansion. He hired four women assistants from the Petrovka and added more than six thousand samples to the seed bank. Among them were the prized seeds he had collected during his years of teaching in Moscow and his expeditions in Iran, Turkmenistan, and the Pamirs.

His lectures were very popular, covering subjects far beyond just agriculture, including the exciting new developments in genetics. In his introductory lecture he told his students that genetics had opened up opportunities in plant breeding that researchers of the past could only dream about. "In the near future man will be able to synthesize forms completely unimaginable in nature," he promised.[9]

Vavilov tried to keep his Saratov botanical community separate from the civil war. He kept politics out of his lectures as best he could. In his spare time, such as it was, he cared for students exhausted by typhus or other diseases. In the absence of a night watchman, Nikolai Ivanovich personally drove the hungry rooks off the experimental fields.[10]

Most of his students were women and one, in particular, attracted his attention. She was a twenty-two-year-old native of Saratov, named Yelena Barulina. A striking, dark-haired young woman with large, slightly sunken gray eyes and a full mouth, she lived in Saratov with her four siblings. Her parents had died young. Her father was of peasant origin and had been a manager in the port. Yelena had graduated from Saratov First Women's Gymnasium in 1913 with a silver medal and was in her third year at the Faculty of Agronomy when Vavilov arrived. As she was a top student, Vavilov would recommend her for postgraduate work and also make her an assistant head of the institute's experimental seed station.

Despite the difficulties of the time and the distraction of a budding romance, Vavilov remained obsessed with his beloved botany. "Work gives a kind of impulse," he wrote to his mentor Regel at the Bureau of Applied Botany in Petrograd. "No matter what you touch, a million things can be done . . . the tasks of botany in Russia and of science in general look broader and of bigger importance to me."[11]

In his first spring in Saratov, in 1918, Nikolai Ivanovich planted more than twelve thousand hybrids of wheat and barley, including the material gathered on his expeditions. Vavilov worked every day on his experimental farm, nothing more really than a field in the black earth, about seven miles from Saratov, where he bred his new crop varieties.

At dawn each morning, Nikolai Ivanovich, freshly shaven and uncommonly cheerful,[12] would be in the fields calling out to any student up early enough to join him, "You are up already?" he would exclaim, and then would dash off into the field. [13]

As more students arrived and began their work, he would suddenly call out, "Everybody, come over here, and hurry up." The young researchers would drop whatever they were doing, run to the spot where he was working, and he would give them a spontaneous lecture on some phenomenon of his plant world, an albino variety, perhaps, or a strange case of gigantism.

The opposing armies of the civil war advanced and retreated across the countryside. Sometimes they trampled over his precious crops so that he had to keep moving his experimental field. He wrote to Regel, "The plot at the farm of the agronomy faculty is now safer because of its distance from the soldiers' camp. So far, there are not many soldiers, but we expect an increase in the near future, and therefore the sowings will be in danger. Last year the sunflower crop was completely destroyed."[14]

At the end of 1918, Nikolai Ivanovich took a rare break to return to Moscow for the birth of his son, Oleg, who was born on November 7, 1918. Vavilov was overjoyed, but also taken somewhat by surprise. He wrote to a woman friend who had just had a child, "You have a multitude of concerns now, I know about it from Yekaterina Nikolayevna's experience. Truly, I used to think that it was an easy thing. But it turned out to be a peculiar occupation that demands a great deal of stamina."[15]

It was a celebration marred by the strain of the upheavals in

the country and inside his own family. His father, Ivan Ilyich, had
been warned that if he did not leave Russia he would be imprisoned. Vavilov wrote to Regel, "Although we have been substan-
tially proletariatized (my father is banished from Moscow), it is
still possible to lodge at the Moscow branch office [of the Bureau
of Applied Botany], where there is a room for you. Meals I cannot
promise."[16] Ivan Ilyich went into voluntarily exile in Bulgaria leaving his wife behind. The family house with the orchard at No. 13
Middle Presnya, where Nikolai Ivanovich and his brother, Sergei,
had grown up, was handed over to the Moscow authorities for use
as a kindergarten. His mother, Alexandra Mikhailovna, moved into
No. 11 where his wife, Katya, and son, Oleg, had been living.

On the day of his father's departure, the family gathered, as they
had when Nikolai left for Persia in 1916, to say goodbye, but unlike
Nikolai, Ivan Ilyich did not go by car. They harnessed the family
horse to a droshky cab, Ivan Ilyich put on his overcoat and his hat, his
suitcases were put in the back, he hugged each member of the family,
and they all wept.[17] When he managed to set up a successful business
abroad, he said, he would call for Alexandra Mikhailovna, but she
wondered, even so, if this was the last time she would see him.

Alexandra Mikhailovna had been resigned to his leaving for many
months, and her anxiety about being left on her own was tempered by
the birth of Oleg. Nikolai Ivanovich declared to his family that the
birth was the best thing that could have happened in his relationship
with Katya and his mother hoped that it was, that somehow Oleg's
arrival would make the couple more compatible. Nikolai would man-
age to persuade Katya to bring Oleg to Saratov where the three of
them would live together for a year, but the marriage was beyond
repair. And his romance with Yelena Barulina grew stronger.

Before returning to Saratov at the end of 1918, Vavilov visited Regel
in Petrograd, the last time he would see his favorite Russian mentor

alive. All his colleagues seemed to be suffering. Lenin had moved the capital to Moscow and there was typhus, cholera, and starvation. In the face of such dreariness, Nikolai Ivanovich always managed to convey hope, even joy. A chance meeting with his friend Konstantin Pangalo, a wheat specialist at Regel's Bureau of Applied Botany, was typical. Pangalo had fallen under a tram and lost his left leg.

"How are things, what's new?" Vavilov demanded when he met Pangalo at work.[18]

"Can't you see I'm a cripple now," Pangalo replied. "I'm not in the ranks of the workers any longer."

"You lost a leg and you're out of the ranks already," scoffed Nikolai, with a friendly laugh, dismissing his friend's amputation as a minor event, easily overcome. "Such a trifle, Konstantin Ivanovich, after all, what are cars for?" And he immediately switched the subject to his new research in Saratov.

"So," he ended, "come to Saratov and you'll see everything and we'll talk."

Pangalo suddenly found his spirits raised.

"After he left I was surprised to notice that I was a different person," Pangalo recalled. "He had given zero attention to my disability, even made a joke about it. He had lifted up my soul."

Nikolai Ivanovich found his friend Regel full of foreboding about the future. Like many in his class—Regel was of high German birth—he saw no real future for himself in the new Russia. Petrograd was in chaos, the city was dying. He had become "Citizen Regel," and, like everyone else, he had to apply for food ration cards, a special pass to avoid the queues at the train station, a pass to walk home after curfew, a pass to bring a sack of food for his family, a certificate of loyalty to the Soviet power, permission to use his skis in the month of February 1918, and a pass to buy tall boots for his field trips.[19] When the bureau needed a small boat for field trips, he had to apply for one official permit for the boat and another permit to row it.

In letters back and forth between Petrograd and Saratov, Regel had engaged Nikolai in a rare personal discussion of the political future of the country, and of Vavilov's own future. Regel had always wanted Vavilov to come to Petrograd and work with him in the bureau, but Vavilov had resisted.

The offer pleased him enormously, but he had felt compelled to stay in Saratov because of a winter planting "that I cannot leave to destiny." He had so many current projects—immunity in plants, hybrids, and some botanical-geographical research—that he felt he could only move to Petrograd "if I could continue with those properly. I am afraid I am too fond of my own freedom in allocating my time."[20]

Both men tried to focus on their science, their letters grimly framing the debates of the day.

"I can see that you also have doubts about the future, Nikolai Ivanovich," Regel wrote in his reply. "What's left for us is to pretend that nothing happened, and to continue work without a second's hesitation, always supported by the reassurance that science is not just apolitical and international, but even trans-planetary."[21]

Vavilov agreed, of course. To him science was everything. But he was still worried. "Events [in Saratov] unfold faster than in Shakespeare's tragedies," he wrote. "If we are alive, if Sodom and Gomorra pass on Petrograd despite its great sins and crimes . . . we will promote the true applied botany."[22]

And Regel saw his own drama. "As far as our intelligentsia is concerned . . . they talk and write straightforwardly, and cleverly, and what they say makes sense. The broad scope of their views is striking. Their erudition is deep, but . . . it's unrealistic. They are hostile to everything concrete. They seek to embrace the un-embraceable. They can't make decisions. They are always talking, smart talk but no action; seeking some eternal compromises and half-measures. And the Messieurs Bolsheviks, they have taken advantage of this situation. . . . We lost, we are defeated. We blundered.

"There is no knowing," Regel went on, "if you, or I, will get out of this chaos alive. It is particularly doubtful for me, because I am not going to make compromises. Until then, let us continue with our peaceful job that is not related to politics in any way. Perhaps we will be lucky to stay unnoticed in the shade, and swim to a dry shore in the end. Perhaps, we'll live to see it."[23] Regel would not witness much more of the Soviet communism he was growing to fear and despise. He caught typhus on a train trip in the summer of 1920 and died, one of three million who succumbed to the disease during the years of the civil war.

In the weeks that Nikolai Ivanovich had been away from Saratov, "everything was nationalized, including apples and watermelons," he wrote to Regel on his return. "There isn't any firewood."[24]

Saratov's ruling Bolsheviks had begun "resettling" the bourgeoisie, confiscating their homes and property and in one case imprisoning them on a ship in the Volga. Others were not so lucky, crammed into prison "20–25 in a cell fit for nine." Under the slogan "Everything for the Front and Everyone for the Front," Saratov came to supply all Red Army operations in the region, draining the city of industrial goods.[25]

"We are not overly optimistic here," Vavilov wrote. "The only way to survive and preserve lives and institutions is with a camel's endurance."

Vavilov was forced by the civil war to transfer all the plantings from the breeding station of the Faculty of Agronomy to a farm even farther from Saratov, but by the standards of those around him he was managing remarkably well. He turned the agricultural institute into the only scientific center in Russia dealing with plant science, breeding, and genetics. The report of the Agricultural Scientific Committee for the period 1918–20 stated: "Of the local establishments only the Saratov branch of applied botany at the university carried out works properly and in full volume . . . including a series of new crop plants . . . and expanding the genetic research on cultivated plants."[26]

Amid all the deprivations and hard work, Nikolai Ivanovich always found time, and perhaps solace, in adding to his expertise. Keen to read the ancient texts on agriculture, he took Latin lessons—from a professor at the university. In those days, it was almost impossible to buy anything for money and Vavilov offered to pay the professor in kind—grain or fresh vegetables from the institute's farm. They had lessons three or four times a week, but, as the professor recalled, they were not normal lessons. Nikolai Ivanovich wanted to learn what the Romans had to say about agriculture, and he insisted on providing a running botanical commentary to the Latin text of Pliny's *Natural History*. "I listened bewildered," the professor would recall. "Others would also have benefited from this commentary, but for once the door to Nikolai Ivanovich's study that was usually open to everyone, was firmly shut during Latin classes."[27]

Vavilov, the eternal student, was also the kind of teacher who inspired extraordinary loyalty in his own students, even under duress. He reported to Regel in 1919 that the year had started "far harder than the previous one. The farm had to be organized from scratch, horses, instruments, forage for the horses, meals for the hired hands, hiring them and equipment—including bridles, nails, firewood. I have become lord and master and I confess my soul is not in it."[28]

Within the year, Regel was dead and Nikolai Ivanovich was offered his job. It was what Regel had wanted. Before he died he had written a glowing recommendation of his protégé to the Commissariat (Soviet equivalent of Ministry) of Agriculture. Nikolai Ivanovich was "the future pride of Russian science," he had said. More than that, as a human being, Vavilov "belongs to a category of people of whom you won't hear a bad word from anybody at all."[29]

Vavilov was appointed director of Regel's bureau, but before he could take up the post he had one important task. He had to deliver a paper on June 4, 1920, at a conference of plant breeders in Saratov. The paper would make him an instant Soviet hero. With the cumbersome title of "The Law of Homologous Series in Hereditary Variability," Vavilov laid out a simple rule for hunters of crop plants: similar features, such as stem size, and leaf size and shape, could be found in the various evolutionary stages of all closely related species, genera, and even families. Vavilov noticed that a trait found in a particular variety of wheat may also be found in barley or oats.

Darwin was the first to note that similar, sometimes even identical, characteristics arise in animals and plants. On the banks of the River Plate between Uruguay and Argentina, Darwin saw bulldog-faced cows that, because of their jaw, resembled certain breeds of dogs and pigs. But Darwin could take his observation no further because genetics, the science of heredity and variation, did not exist. In the same way that the Russian chemist Dmitry Mendeleev had brought order out of chaos half a century earlier to the elements in drawing up the Periodic Table, enabling chemists to predict the existence of undiscovered elements, Vavilov was trying to bring order to the organic world, guiding plant hunters to look for "missing" forms, or "gaps" in a species. Until now, plant hunters had collected varieties of cultivated plants more or less at random.

The local Saratov newspaper trumpeted "Professor Vavilov's Discovery," and the report said the Saratov authorities were so impressed that they had decided to publish Vavilov's works and to provide him with a well-equipped state farm for large-scale experiments—even to fund scientific expeditions to other countries.[30] The lecture hall was packed for the occasion, standing room only. Well-known biologists came from Moscow and the whole of the teaching staff of Saratov University attended. When Vavilov had finished, the hall remained silent for a moment and then erupted into applause. He was given a standing ovation, one professor declaring, "The biologists congratulate their Mendeleev."[31]

The Russian agronomist Nikolai Tulaikov, who participated in the meeting on behalf of the Experimental Division of the Commissariat of Agriculture, the leading Soviet agency covering research into agronomy, reported on Vavilov's outstanding work and called him "the pride of Soviet science."

Vavilov now saw before him great expeditions to Asia, Africa, and the Americas to fill in the gaps in a new classification of plant life on earth. But the exhilaration of the comparison to Mendeleev would dissipate. There were, of course, so many more plants than there were elements, and even more variations in the plants than one could possibly fit onto a chart like Mendeleev's periodic table. Also, Vavilov had proposed his law prior to the theory of chemical and radiation-induced mutation and what this theory would mean for evolution and efforts at botanical classifications.[32]

For the moment, Vavilov availed himself of an equally intriguing and rewarding botanical experience at home in Russia, a visit to Ivan Michurin.

CHAPTER 7

The Gardener of Kozlov

Here for ages gardeners drudge
with glass topped frames and heaps of dung,
but in my hand
on roots of dill
six times a year pineapples grow.

—VLADIMIR MAYAKOVSKY, 1918[1]

In his satirical poem about the new generation of horticultural-
ists, the Bolshevik poet Vladimir Mayakovsky mocked Russia's
most famous amateur gardener, Ivan Vladimirovich Michurin. The
slightly batty son of an impoverished nobleman, Michurin and his
green thumb became a much needed figure of fun in the socialist
utopian dream, a goofy sort who thought it might be possible to
"grow pears on pussy willow."

In reality, Michurin, who was sixty-five in 1920, was a model
representative of the old school of "artisan" gardeners. He knew
nothing of Mendel and genetics and, as he might have explained on
a tour of his orchard, he was too old to begin to understand such
highbrow concepts. His method of breeding new varieties was tried

and true. He relied essentially on the same criteria farmers had used over the centuries: a knack for spotting a sturdy plant and knowing intuitively when to select the right seedling, when to be ruthless in casting out the weaker ones, and how to nurture the stronger ones to maturity. He employed the time-honored gardener's trick of grafting a cutting of one plant onto another thus producing a new shoot with tissues from both plants. He was a leader in Russia's gardening subculture that flourished in the orchards and fields far from Moscow and Petrograd. Over the years, the more enlightened peasants had coaxed better yields from the crops on their farms. However, none was quite as adept, as hardworking or inventive, as Michurin. His magic was legendary, especially when it came to fruits that were such a luxury in Russia's north.

Michurin lived in Kozlov about two hundred miles south of Moscow where the winters can be unforgivingly cold. The fruit varieties that survived were necessarily hardy but lacked the succulence of those grown in the warmer republics of the Crimea and the Far East. Michurin's lifelong goal was to crossbreed both types to produce a better fruit for the northern territories. He would eventually produce some 350 varieties of apples, pears, plums, apricots, peaches, and grapes. Michurin, whose name is barely known now outside Russia, was a gardener-artist in the orchard, a worthy rival of his American counterpart, Luther Burbank, the flamboyant gardener of Santa Rosa, California.

Despite his success, however, few of Michurin's varieties were used by Russian farmers, who saw him as a horticultural wizard with special powers that were necessary to achieve his success. If they tried to grow his varieties, they imagined they would fail, so most didn't even try. In 1920, he was known only vaguely to Soviet agricultural officials, most of whom regarded him as something of a crank. The first Soviet administrator to recognize Michurin's talent was Nikolai Ivanovich. Curious, as always, to find out about any botanical success on Russian soil, Vavilov yearned to visit the inventive fruit grower.

The first opportunity came as the botanists who had attended the congress in Saratov were going home. One of the professors, who knew Michurin, invited the participants to visit his orchards. Vavilov readily accepted: if a breeder had good seeds, whether he was a trained scientist or not, Vavilov wanted to have a look.

The meeting was a great success, despite their widely different backgrounds and the fact that they were a generation apart. Vavilov was the explorer and well-educated scientist on the cutting edge of international research into plant genetics, about to become director of plant breeding in Russia. Michurin was the amateur gardener, steeped in folklore, bound by poverty to a tiny farm in one of the poorest regions of Russia. But the two men had a common, overweening purpose: to collect the best varieties of plants from wherever they could find them and improve the food that people eat. And they went about their business with such gusto they left other mortals standing in amazement. Vavilov had come to Kozlov to see what, if anything, Michurin could contribute to the great task of transforming Russian agriculture.

The Michurin family fruit farm had failed in the 1870s, long before the revolution. Kozlov, as far north as Saskatchewan, was a difficult place to grow fruit, but that was only one reason. Ivan's mother had died of tuberculosis and his father was in and out of the madhouse. Michurin, who was born in 1855, grew up poor, completed only one year at the gymnasium, worked as a railroad clerk and a watch repairer, and married a woman from a lower class.

For the noble Michurins, this should have been the end of the road, but in the 1870s Ivan was determined to get back into the fruit growing business. He bought a thirty-four-acre plot in Kozlov with borrowed money and started planting trees. Somehow he managed to bring in varieties from all over Russia, especially from the Crimea. By the turn of the century he was still poor, but becoming known because of his new adopted varieties.

Being of the old school, when Ivan wasn't grafting cuttings, he was growing fruit trees from seed and crossbreeding them. It is a tricky business at the best of times because the results are usually wildings, rangy specimens of inferior quality. But after some success, he tried to interest the tsarist agriculture ministry in his work: as a "true Russian man" he considered it his "sacred duty to offer his humble labor for the good of the fatherland."[2] Two and a half years later the ministry replied, offering some small assistance, so paltry that Michurin considered it beneath him, and he declined. Still, Michurin kept producing varieties of apples, pears, peaches, cherries, and apricots that were both hardy and delicious and that attracted breeders from Europe and America.

In 1911, one of America's most famous plant hunters, Frank Meyer, visited Michurin on assignment from the U.S. Department of Agriculture. He reported to Washington that Michurin had some "extremely valuable" material that represented years of patient work. Michurin explained how difficult his task had been. For example, in an attempt to create a hardy peach, he planted peach kernels sent in from all regions of Russia. He had three thousand young trees at one point, but only fifteen survived the winter. Then these survivors were killed off by some kind of bark rot.

Under orders from Washington, Meyer offered to buy samples. But Michurin would only give them away, and then only some of them; not his prize possessions.[3] Meyer concluded he was a "peculiar sort of man," and no deal was struck. Michurin was once again impoverished and without prospects or support from his own government, or anybody else.

Nikolai Ivanovich was shocked at the poor conditions—"the squalid state of the experimental farm; the wretched little wooden cottage in which one of the most remarkable plant breeders of our times lived and worked."[4] Michurin himself was gaunt, skinny with sunken eyes and an unkempt beard. But his spirit was strong and his research, though primitive and unscientific, appealed to Vavilov. So, the lofty government plant breeder decided to help the impoverished farmer. It was the typical gesture of generosity from a man who was already thinking of how to dispense public funds even before he had taken control of them.

It is worth restating what in hindsight now seems obvious. Vavilov was the kind of scientific purist who always tried to keep his mind open. The scientist who is so certain of his own position can easily fail to see the larger possibilities. Perhaps the secretive and curmudgeonly Michurin was onto something that the geneticists, in their headlong pursuit of genetic theory, had overlooked. Perhaps Michurin had devised special techniques of propagation that could also be useful to the experimental breeding stations in other parts of Russia. Perhaps his methods could help feed more people, or, and this was always Vavilov's hope, could shorten the time it took to turn his precious seeds into new and better varieties.

Vavilov would start a lively exchange of letters and plant specimens with Michurin. Once he was installed in Petrograd, Nikolai Ivanovich would send a specialist to evaluate Michurin's work, and he would ask the Commissariat of Agriculture to celebrate Michurin's forty-five years as a breeder with some kind of official recognition. The government would eventually grant Michurin ownership of the land he farmed for the rest of his life, tax-free, half a million rubles, and a promise to publish his written works, such as they were, with Vavilov as editor.

Michurin would be eternally grateful: not only was he comfortable, he could spend all his time in the orchard. Michurin would tell a friend, "Vavilov is an outstanding scientist, a brilliant mind . . . he travels all over the world collecting the plants we need. . . . And it's quite amazing—he knows about a dozen languages. . . . And I must say he is very sympathetic to the work we are doing."[5]

As a member of the scientific elite, Vavilov was indeed exceptional in the way he helped colleagues, young and old, whose work he felt merited his support. As long as they demonstrated some of his own legendary qualities of energy, enthusiasm, inventiveness, and hard work, he was prepared to back them. For most, he insisted on strict adherence to scientific analysis and verification. He invited debate, but not quarrels, alternative theories but not warfare. Over time, he tried to get Michurin to understand Mendelian genetics, but Michurin either could not or did not want to learn the new science. He had been following his own primitive methods for more than half a century and he had no desire to change. Vavilov made an exception for the old gardener of Kozlov. He was, after all, of another generation.

Before moving to Petrograd, Vavilov returned briefly to Saratov for one last botanical expedition—four hundred miles by riverboat down the lower Volga to Astrakhan on the Caspian Sea. The official aim was to study field cultures of the southeast and in the Volga delta. The group included two professors from Saratov University and an expert in melons. But there were also two research students, Gali Popova and Nikolai Ivanovich's secret love, Yelena Barulina. The only record of this trip is from Popova, but one can imagine how romantic it was for the young Yelena to be whisked out of war-torn Saratov to spend the next four weeks in the close confines of a riverboat with the man she had come to love but, until then, had only been able to admire from a respectful distance. Popova recalled several of the idyllic settings along the way, beginning with the departure from Saratov. "The Volga was calm as a mirror, the steamboat was late and all members of the expedition surrounded Nikolai Ivanovich and listened to his engaging stories of plant-hunting in Iran and the Pamirs. The sun had already gone down when the lights of the boat appeared and, against all our expectations, they put us all into cabins."[6]

In Astrakhan, Vavilov took them through the market to check the local varieties of fruits, grapes, apples, pears, and watermelons.

Then they explored the delta in small motorboats, where they found lotus bushes and water nuts. "Nikolai Ivanovich was so excited his boat was filling up fast with the nuts . . . suddenly a miraculous sight appeared before us: beautiful pink lotus flowers stood as if on legs among the cattails. All the boats headed towards them and we took samples for the herbarium."[7]

At the end of the voyage, Nikolai Ivanovich asked Yelena Barulina to accompany him to Petrograd where he would give her a job in the Bureau of Applied Botany. She was enchanted by the proposal, but overwhelmed. She could only bring herself to promise to write to him. The difficulties and uncertainties facing Vavilov were reflected in a letter to his friend, the geography professor Lev Berg. Vavilov announced his intended arrival in Petrograd that fall, adding, "Moving over together with the laboratory is hard. This migration from the southeast [where there was generally more food than in Petrograd] appears unnatural to many, but everything is complicated, and bread is not man's only food."[8]

CHAPTER 8

Lenochka

Nikolai Ivanovich arrived in Moscow in the middle of September 1920, en route to Petrograd to take up his new job. The young Yelena Barulina had indeed been overwhelmed by the Volga trip and his offer to join him. She told him that she needed time to consider his proposal.

Yelena Barulina's first letter came soon after Nikolai Ivanovich arrived in Moscow in September 1920, where he had stopped to see Katya and his son, Oleg, now almost two years old. Yelena told him of her yearning to join him, but also of her doubts. The difference between their two lives was overpowering to her. He was the dashing young professor about to take a top government job. She was a research graduate, an uncertain twenty-four-year-old. He had already traveled widely. She was a provincial girl who had never left Saratov, her home, or her close family. She had a steady but unromantic relationship with a local man. Nikolai was married with a son, and they had never talked about his other life. Now, given the comfort of distance, they could write about this hard reality. She needed to know about his relationship with Katya, his real feelings toward his wife and his pull toward his son.

One can imagine the agonies for Yelena. She was living with her four siblings in a modest home in central Saratov, where her comfortable world had been collapsing around her, closing in on her social class, the middle class. Prices had shot up 900 percent since the revolution. The ruble was worthless. The bread ration had been cut to one quarter pound per person per day. Hunters roamed the forest in search of game. There were epidemics of typhoid fever, cholera, dysentery, the Spanish flu, measles, and diphtheria, and medicines were running out. Hospital patients were not being fed. Salaries were not being paid. Able men were dragooned by the authorities into cutting hay and firewood on the outskirts of the city. Red Army soldiers, returning from the front, were sleeping in the streets. Houses were "nationalized" at random and turned into barracks. A great fire on the outskirts of the city spread rapidly, destroying houses where 25,000 people had lived, about one eighth of the city's population. Surely, Yelena must have thought, life with a brilliant professor in Petrograd, the gleaming city on the Neva, the center of Russia's intelligentsia, must be brighter.[1] But what if he left her there alone, what if his wife and child pushed her out? What if he found another researcher?

When he read that first letter in Moscow, Nikolai Ivanovich waited until Katya and his mother had gone to bed before sitting down at his absent father's desk to write his reply. It was not easy for him to express his emotions. He was a proud man of science, a man who dealt in observable facts and theories, and as a child he had become accustomed to suppressing his inner feelings. A stern, authoritarian father and a mother whose life was governed by tradition and the church had not given him room to express his strongest emotions. Yet, unlike his stiff travelogues to Katya, Nikolai Ivanovich managed to write a full confession of his love for Yelena.

"Darling Lena, I got your first letter. I needed it so much. . . . We need to know each other closely. Of course, you must tell me about everything that binds you. . . . I came to love you without knowing of your personal life. From the snapshots that I saw I reconstructed your life in my mind. I knew intuitively that I loved you. . . . Everything that I learned from your letter made you even dearer to me, darling Lenochka. . . . After your reply, my strong desire is for our love to be firm and strong."[2]

For a long time, he said, he had been "prevented from making my confession." He had been so wrapped up in his scientific work he had not analyzed his own feelings objectively. He had known for some time that he did not want to spend his life with Katya, but it was only now, with Yelena, that he was learning why and what was important to him.

"To love means to be willing to see your beloved at all times, to wish to share your emotions, to live in unison, and, if possible, to work together." There were "no such things" with Katya, he wrote.

He had known Katya for a long time, he explained, since student years, but they had never been close. She was the cleverest and best-educated student at the Petrovka. She was respected by students and professors alike. But under all that brilliance was a difficult side, a sourness that Vavilov's buoyancy could not lift. He tried to explain it to Yelena.

"An attempt was made to follow a single path together, but nothing came out of it. Besides, Yekaterina Nikolayevna's bad temper was always there. Now the only thing that links us is the son, who is impossible not to love. There are many of my features in him, and I would like to pass on to him all the best I can. I would very much like it if he were dear to you, too."

Nikolai Ivanovich wanted Yelena to know that his family would be difficult. His mother would press him to stay with Katya for the sake of the child. So would his brother Sergei, who had always respected Katya.

"My mother is a kind, simple woman, almost elderly now . . . and she is going to be very stern because of Oleg. . . . I love my brother

Sergei, although we are not too close. He is very gifted and he will likely become a prominent physicist. He has a lot of respect for Yekaterina Nikolayevna and certainly he won't be on my side, at least at the beginning. My family adheres to old traditions of which I disapprove."

His older sister, Alexandra, would be more sympathetic, he said; he had watched her suffer from his parents' disapproval after she was abandoned by her husband and left with two small children. Alexandra had "been through a lot, looks at it simply and calmly, the way I would do, too. . . . I would like it, after everybody comes to terms with everything, and surely this is going to happen, to be close to your family if you find it necessary."

But he did not want to make excuses for himself. "Everyone is responsible for his own life. On November 25, I am going to turn 33. Here I am going to be short and frank. I am writing to encourage you to do the same. I am interested to know everything about you."

And he wanted her to know a little about himself, his "essence," especially his devotion to science, and when she understood that devotion, he was sure that things would work out.

"I am not a pessimist, rather I'm an optimist. My youth was not as full as I would like it to be."

At four in the morning, he ended his letter with a strong commitment.

"It's time to finish and force myself to sleep. My darling, beloved Lena, I want us to understand each other in everything, for our love to be strong. I want us to be friends who share everything, grief and joy. I am endlessly happy that we are going to work together. Dearest beautiful Lenochka, I want you to be happy. We are right in what we are doing. All complicated things will turn simple. All difficulties can be overcome. There's so much work to do. Everything speaks in favor of moving to Petrograd. There we will build our life together. Write to me, sweetheart, I am longing to see you. Your N.V."

But he had not quite finished. Before putting the letter into an

envelope he felt that he had to add something—about his passion for
science and how that would fit into his relationship with her.

"There is so much I'd like to do," he wrote in a postscript. "You
know my plans a little, they are not fully formed. You will walk
along with me, and I am delighted to have my closest dearest friend.
I got used to linking my life with science. I shall put all my efforts to
please you (be very strict). Sometimes, like now, I feel I am capable
of achieving something. Happiness gives strength. It's been a long
time since I was so happy. Your N.V."[3]

Without her help, he seemed to be saying, his passion for science
would continue to dominate his life, as it had done with Katya; his
love for Yelena would always take second place to science—unless
she was "strict" about preventing it. In Yelena he had found romantic love for the first time, he had never been so happy, but, knowing
his obsession with science, he was afraid he would not be able to
keep her.

A few weeks later, at the end of October 1920, Nikolai Ivanovich
took the train to Petrograd to begin the task of taking over the small
Bureau of Applied Botany, then housed in six apartments in a building in the center of the city. Vavilov had already been warned what
to expect. Before he died Regel had told him that there were "big
difficulties with housing, furniture and food for new arrivals . . . tables and stools are not delivered, there are no bookcases, there is no
money."[4] But what Vavilov saw made him wonder if he could ever
complete his task.

Petrograd was closing down—factories, shops, universities,
offices, schools, and hospitals were abandoned. In an attempt to ease
the burden on the citizens, the local soviet abolished rent for apartments. Water and electricity, when available, were free, and fares on
public transport were eliminated, a small consolation since the trol-
leys seldom ran. The city-run steam baths were also free, but had

long been closed because the water pipes had frozen. And there was almost no food. The writer Maxim Gorky wrote to his friend H. G. Wells, "There is practically no food and it can be said without exaggeration that there will soon be famine in Petrograd. . . . I just can't imagine how our scholars are going to survive."[5] In an obituary for his friend Regel, Vavilov would write about his own concerns about the fate of Russian science. "With every day that passes the ranks of Russian scientists grow thinner and thinner and the fate of Russian science lies in the balance. Replacements are many, but few of them are real and one is horrified for the fate of Russian science because many are called but few are chosen."[6]

On the evening of November 5, 1920, Nikolai Ivanovich sat in his overcoat at Robert Regel's desk trying to keep warm in the unheated office and listening to a report on the burdens on those still trying to survive in the city; the cold and the hunger and the lines for food were now part of the daily routine for everyone in the former tsarist capital, for workers, party members, and the former bourgeoisie. It was after midnight, when the staff had long gone home.

"One sad thought follows another," he began his next letter to Lenochka. "New life needs to be breathed into everything, as there is almost no life left here; it is as if things were, if not dead, then seriously ill, paralyzed. Everything needs to be rebuilt. Only the books and good traditions remain undying."[7]

Nikolai was especially worried that he would not be able to look after the forty people on the staff of the bureau now in his care. "Many of them are good and excellent workers. Several are getting ready to leave because of the deprivations. They have been waiting for changes for the better on my arrival." He was ready, even eager, to take on the challenge, but once again this outwardly robust and confident young man expressed his inner doubts to his Lenochka.

"Dear friend, I fear that I will not be able to cope with everything. It doesn't depend on one person. Rations, firewood, salaries, clothes. I am not afraid of anything, and difficulties long ago became a challenge. But I fear not for myself, but for the institution and fellow researchers. It's not just a matter of organizing work productively, which I can do, but of putting the personal lives of many in order. It is all more difficult than it appeared from afar."

Yelena still had her doubts and such news could not have been encouraging. As best he could, Nikolai Ivanovich tried to allay her anxieties and he offered her the highest compliment he could give—his love for her was as strong as his love for science.

"My dear friend," he wrote again three weeks later, "you are plagued with doubts as to whether the emotion, the burst of passion will simply pass. Dear friend, I don't know how to convince you, or what objective evidence I can offer you to prove that this is not the case. I myself would like to step aside and subject myself to unsparing analysis.

"It seems to me that despite my tendency to bursts of passion and emotion, I am nevertheless a very constant and dependable person. I take love too seriously. I really have a profound faith in science, in which I find both purpose and life. And I am quite ready to give my life for the smallest thing in science. . . . And that is why, Lena, simply because I am a faithful son of science, I cannot, within myself, allow myself to treat matters of the heart lightly. . . .

"I do not have any great demand for comfort. True, I am not used to doing everything for myself, but I can do it when it is absolutely necessary. We won't have any arguments about that, I am sure. I really don't know what we shall disagree about. Our life together should be very beautiful, both emotionally and in our everyday work. And since you agree on that; it seems to that our union will be strong and lasting."[8]

In his relationship with Katya, he wrote, he had found himself, like Dante in *The Inferno*, "*Nel mezzo del camin di nostra vita*—halfway along life's path . . . I strayed into a dark forest, a dense forest."[9] He admitted that there had been frequent "near-quarrels" with Katya, but he insisted that he had done nothing to make them worse, and that he was normally tolerant and easy to get along with. The future was bright. He had turned thirty-three the day before, he noted again, and it was time for a change.

"Now I have to get out of that forest. And I believe that we will get out of it together. It is a difficult forest, but is there any forest which does not have a way out?"

Yelena would make the next move by agreeing to come to Petrograd, but their relationship was to be kept a secret even from friends for another six years.

CHAPTER 9

Petrograd: City of Ravens

Nikolai Ivanovich moved permanently to Petrograd with a group of his Saratov researchers in the spring of 1921, just in time for the planting season. With the end of the civil war, it was a critical time for the revolution. There were mass demonstrations and strikes in Petrograd and other cities, as people demanded changes in the civil war policy of confiscating surplus grain. A revolt at the Kronstadt naval base, an island off Petrograd, would be brutally repressed. The Bolsheviks were forced to change course. Lenin's New Economic Policy legalized free trade in food and consumer goods and brought a relative peace to the rural areas. But even that reversal was too late to prevent the terrible famine of 1921–22.

The train from Moscow, straining under its load of freight cars, pulled into the station at Petrograd on March 5, 1921. When the door of a cattle car slid open, Nikolai Vavilov and ten of his young plant researchers from Saratov jumped onto the platform. They were mostly women in long black woolen dresses and dark gray coats. Yelena Barulina, his beloved Lenochka, was not among them. She would arrive in two months, with another dozen of his workers from Saratov. This first group carried their scant belongings in parcels and they were exhausted, hungry, and cold. The food they had packed when they left Saratov in early February had run out soon after leaving Moscow. As they moved north, the temperature dropped until it was again below zero, endurable only if they huddled together like farm animals. On the way they had stopped countless times at stations where wrecked locomotives and idle carriages lay as evidence of the national chaos.

Even so, they were in high spirits, infected by Vavilov's enthusiasm and eager to join what was left of the Petrograd intelligentsia. To pass the time, they sang songs, including one they had made up about their dear Professor Nikolai Ivanovich. The song told how he would take the Petrograd scientific community by storm; how the name of Nevsky Prospekt—Petrograd's most famous street, renamed after the revolution to Avenue of 25 October—would soon be changed again, this time to the Avenue of Homologous Lines—to commemorate Nikolai Ivanovich's acclaimed theoretical work on variation in plants.

As they pulled into Petrograd they saw what Vavilov had witnessed on his short visit in 1920: the cradle of the revolution was disintegrating. The intelligentsia's dream of a bourgeois paradise on the Neva, an open city free of the Bolsheviks who had decamped to Moscow, was a mere dream. Petrograd, once the mighty St. Petersburg, was at a third of its prerevolutionary level of 2.3 million people. The atmosphere of quiet despair would be caught in the lithographs of Mstislav Dobuzhinsky, who found the city "dying before my very eyes with a death of incredible beauty, and I tried to capture as best as I could its terrible, deserted, and wounded look."[1] The avant-garde artist Yuri Annenkov, who would emigrate to France, recalled "endless hungry lines, queues in front of empty 'produce distributors,' an epic era of rotten, frozen offal, moldy bread crusts, and incredible substitutes."[2] Anna

Akhmatova, Petrograd's fashionable poet, wrote about the feeling of hopelessness:

> *In the West the winter sun still shines,*
> *And rooftops still glitter in its rays,*
> *While here death covers houses with crosses,*
> *And calls the ravens, and the ravens come.*[3]

The Petrograd intelligentsia—the educated middle class in Western terms—had emerged exhausted from the anarchy of the civil war and fearful of the direction of Soviet power. Lenin, who had moved the capital to Moscow in 1918, despised the intelligentsia and the institutions of old Russia, and he meant to replace them as rapidly as possible. In 1919, he wrote to his friend Maxim Gorky, "The intellectual forces of the workers and peasants are growing and gaining strength in the struggle to overthrow the bourgeoisie and its henchmen, the intellectual lackeys of capital, who imagine they are the brains of the nation. Actually, they are not brains, but shit."[4]

The Bolshevik slogan was "Knowledge and Education for the Masses." In the Bolshevik mind, many "professors" and "academicians" were automatically considered enemies of Soviet power,[5] and many had been harassed, arrested, sentenced, and even executed in the first waves of "Red terror" during the civil war. Among the early victims was Nikolai Koltsov, a famous cytologist and biologist. He was arrested for his past membership in the Kadets, the Constitutional Democrats, the most influential liberal party before the Bolshevik takeover, but his life was saved by the intervention of Gorky, who appealed directly to Lenin. The anarchy and instability of the civil war forced hundreds of scientists to flee abroad, an exodus that strengthened Europe and America as much as it weakened Russia. Among the famous names in science was Igor Sikorsky, the aircraft engineer. He settled in the United States. A young chemist, George

Kistiakowsky, also went to America, where many years later he would become a science adviser to President Dwight Eisenhower. The biologist Vladimir Korenchevsky took his talents to Britain.

Many artists, writers, and musicians fled to the West as well, among them Stravinsky, Prokofiev, Rachmaninoff, Chagall, Kandinsky, Nabokov, Pavlova, and Balanchine. Even Maxim Gorky had retreated to Europe during this dangerous time.[6]

Despite his loathing for the intelligentsia, Lenin understood that the country needed the "bourgeois specialists" who had been trained under the tsar and were working in a few hundred institutes and research labs, many of them in Petrograd. These highly trained scientists and engineers could help build socialism in Russia and put the country back to work. Lenin was prepared to indulge them, for now. Those specialists, like Vavilov, who chose to cooperate with the Bolsheviks would receive very generous support considering the meager resources available at the time. Several large new research institutes were established including a permanent Atomic Commission with physicists who would later play important roles in the Soviet atomic bomb project. This government-sponsored initiative was an innovation in international science; in Europe and America research was carried out at universities uncoordinated by central government.

The specialists would be able to move into much grander accommodations than they had experienced under the tsar. In Petrograd, they had access to the abandoned palaces and houses of the aristocracy who had fled the country. The specialists were also given extra privileges. In December 1919, the Council of People's Commissars, the highest governmental body under the Politburo (the political bureau of the Central Committee), had issued a decree providing the specialists with extra rations of food and fuel, in the hope of preventing them from emigrating and thereby maintaining a solid scientific base for rapid industrial and agricultural expansion at home.[7] The young Vavilov had no thoughts of leaving his Mother Russia;

he was only too eager to accept the opportunity the Bolsheviks were offering.

On the Petrograd station platform, Nikolai Ivanovich, irrepressible as ever, rallied his research team to the immediate task of unloading his botanical treasures. For him, science continued in war, in famine, or in plenty. In an adjoining freight car he had packed his seed collection, including the rare wheat and rye he had so carefully brought from Iran and the Pamirs. There were books, scientific journals, microscopes, and other equipment from the Saratov laboratory, a heavy load for his band of helpers. Somehow, out of nowhere, Vavilov the magician produced several horses and wagons to carry his bounty carefully down Nevsky Prospekt. They moved slowly past the proud reminders of another time—the grand eighteenth- and nineteenth-century facades behind which bankers and the nobility had conducted their commerce—to the modest offices of the Bureau of Applied Botany.

The offices brought little relief. As they walked through the door they saw what one assistant would describe as "a picture of almost complete destruction, as if there had been an enemy invasion. . . . The rooms were freezing, the central heating pipes had burst, the bulk of the seeds had been eaten by starving people, there was dust and dirt everywhere, and only here and there signs of life; a few lonely and dejected-looking technicians lacking any leadership."[8]

The academic institutes were only now beginning to appreciate their new privileges. Several institutes had already moved into new premises, taking over abandoned tsarist palaces. The Geographic Institute had transferred its maps and data into the palace of the Grand Duke Alexei Alexandrovich.[9]

After a week in Petrograd, Nikolai Ivanovich wrote to a colleague in Saratov, "There are millions of troubles. We are fighting against the cold at home, and for furniture, flats and food. . . . I must confess that it is quite a problem to arrange a new laboratory and an experimental station as well as to settle 60 employees. I am accumulating patience and persistence.

"About three weeks will pass before we settle everything, and then the planting season will approach. We shall need horses, tools, workers. . . . [It] will be much more difficult here than in Saratov."[10] The shortage of food was more acute in Petrograd than elsewhere. Vavilov wrote to a colleague, "Such a move from the southeast would seem unnatural to many . . . but 'man doth not live by bread alone.'"[11]

Nikolai Ivanovich coped—somehow—with the needs of his researchers even though he would go for days without finding enough food. The wife of another Petrograd professor recalled how Vavilov turned up one evening and brought a small bag of millet and a tiny piece of bacon, which he asked her to cook. He later confessed it was the first hot meal he had eaten in a week.[12]

Shortly after his arrival, Vavilov found new accommodations for the Bureau of Applied Botany in the former tsarist Ministry of Agriculture, a splendid palace built in the nineteenth century for a nobleman in St. Isaac's Square. This was one of the city's most prestigious squares where the tsars had been crowned in the golden domed cathedral. In addition, he found a space for his experimental farm in the village of Pushkin, about forty miles from Petrograd in the grounds of the tsar's summer palace of Tsarskoye Selo. His headquarters was in a replica of an English country house, with exposed beams, red roof tiles unknown in Russia, and a magnificent oak staircase. The house had been sent by sea, piece by piece, tile by tile—by Queen Victoria to her godson, the ex–Grand Duke Boris Vladimirovich. Attached to the house were stables and a large luxurious greenhouse where Vavilov immediately started to grow his seedlings. He wrote to his friend William Bateson in England, "Much of my time is taken up with the organization of our new experimental station in the environs of Petrograd. You will be rather astonished to hear that the country house we live in was presented many years ago by the late Queen Victoria to her godson. . . . The country seat is very charming. . . . During the past four years, unfortunately, the main buildings were occupied by comrades."[13]

As Vavilov was getting established in Petrograd the Tenth Soviet Communist Congress opened in Moscow. It would be a turning point in the early history of the Soviet state and, for a brief period, the change would improve agricultural research and food distribution. Proposals for a compromise economic policy had been under discussion in the party's Central Committee since the beginning of February, but any retreat from the road to socialism had been opposed by Trotsky and other left-leaning Bolsheviks. Lenin now proposed the New Economic Policy (soon known simply as the NEP). Under this new policy, the state kept its exclusive control on the "commanding heights"—on finance, large and medium industry, modern transportation, foreign trade, and wholesale commerce. Private enterprise was allowed to creep in at the bottom, including the production and distribution of food. The forced requisition of food surpluses during the civil war was replaced by a proportional tax, payable in kind, particularly for staples like grain or potatoes. The state grain monopoly was thus abolished and peasants could suddenly keep and sell whatever produce they had left over.

The introduction of the gold-supported ruble quickly stimulated economic development and the launch of a scientific and cultural revolution. Lenin personally intervened to fund scientific projects and institutes, and international scientific cooperation, abandoned during the wars, was restored. This was a terrific boost for Vavilov personally.

For the next eight years scientists were allowed to go abroad, with permission of course, and funds were allocated for Vavilov's botanical expeditions. In 1920, only ten scientists from the Academy of Sciences were sent abroad; two years later seventeen would go, and in 1926, forty-four. There was a noticeable increase in government funds for scientific research and in the number of research publications.

Relatively well-off families could now find produce that had disappeared during the civil war—cauliflower and Brussels sprouts, asparagus and shallots, carrots and pods of green peas. German and French confectionaries opened selling éclairs and pralines and forgotten pastries, Vienna rolls, petit fours, and cream horns.[14]

The NEP reawakened some of the old entrepreneurial instincts and improved the lives of many ordinary Russians. But it could not be a cure-all. The sudden burst of free trade in food could not avert the looming famine that would be the worst ever experienced in Russian history. The harvest of 1918–19 had been poor. In 1921, when a good year seemed possible at the end of the civil war, the crop was devastated by drought. The wheat withered throughout the long summer and the harvest produced only 36.2 million metric tons, half the normal prewar level. Lenin would swallow his pride and accept international food aid, especially from America. The largesse from a nation that would not become the enemy for another generation also helped Vavilov. The new Soviet state sent him on a mission to the United States to examine American agriculture and to buy seeds suited to Russia's rich and underproductive farmland.

CHAPTER 10

Ingots of Platinum

The theory of genetics was developed in Europe, but by the 1920s American scientists had not only caught up but overtaken their European colleagues. One of the chief centers of genetics research was Thomas Hunt Morgan's fruit fly lab at Columbia University. Morgan would discover numerous cases of mutations and he confirmed that chromosomes were the carriers of heredity. At the same time, the U.S. experimental crop breeding stations were developing new varieties of beans, wheat, cotton, maize, tobacco, and potatoes according to Mendel's laws.

As the Russian famine was taking its terrible toll, Lenin was forced to think about a long-term plan to improve Russian agriculture. On a train ride from Moscow to Petrograd where he could see his starving citizens, pathetic, emaciated figures, Lenin is supposed to have declared, "The famine to prevent is the next one. And the time to begin is now."[1]

Lenin was ready to spend lavishly on science, and the key to Soviet agriculture, as he saw it, was exactly what Vavilov had in mind: a scientifically based system for testing and improving new varieties. With Lenin's personal support, Vavilov would turn the small, underfunded Bureau of Applied Botany into a vast plant breeding empire.

Soviet science in this period was organized on a system of personal patronage. If a scientist needed support and the funds were not forthcoming from the relevant government agency—in Vavilov's case the Commissariat of Agriculture—then the scientist could appeal directly to the head of the agency or to someone of influence in the government. The entrepreneur in Nikolai Ivanovich quickly understood the system. He would strike up a rapport with Lenin's former personal secretary, Nikolai Gorbunov, the executive secretary of the Council of People's Commissars, the head government agency that was above all the other commissariats, including agriculture. Under the patronage system, scientists like Vavilov became governmental "trustees," experts in their discipline who could wield great influence. In this way in the 1920s, Vavilov would affect almost the entire institutional development of plant science.[2] And it was through this system that Vavilov set up his first major plant hunting expeditions.

Lenin's thoughts on the matter of improving Soviet agriculture were influenced by an American popular science writer, William Harwood, whose 1906 book, *The New Earth*, was translated into Russian three years later. The book was a romantic, populist view of how the practical farmers of America had taken the science of plant breeding out of the labs of ivory-tower academics and into the fields. The model breeder of the book was Luther Burbank, the American horticulturist famous for many varieties of vegetables, fruits, nuts, and berries. Burbank, who had no formal education and was really no more than a clever gardener, was portrayed as the "most practical of all men" and a "pre-eminent man of science," although there was little discussion of science in Harwood's book.

Here, right in the heart of capitalism, was an example of the Marxist view of science that stressed the need for the unity of theory and practice. Lenin understood that any plan to breed new varieties would take many years, and as a start in 1921 he issued a decree, "On Seed Production," that required seed testing in special research stations.

The immediate task, however, was new seed to replace the stocks that had been eaten by starving citizens, and in July 1921 the Commissariat of Agriculture ordered Nikolai Ivanovich to go to America on an urgent mission: buy the best seeds he could find for the next spring sowing. For all the urgency, the trip sounded like a perfect assignment for the adventurous young agronomist; a transatlantic voyage with a brief to travel anywhere across the United States. It would give him a chance to see, firsthand, the American plant breeding system he had so greatly admired from afar. But getting there was not so easy. There were no formal diplomatic relations between the two countries and his difficulties began in Moscow. When Soviet citizens were leaving on official business, even one sanctioned by Lenin himself, they had to get permission from a bewildering array of government departments, including the Cheka, the state security police. Nikolai Ivanovich wrote Lenochka about his frustrations.

"If I had known before how much trouble America would be, I would, perhaps, have abstained from the enterprise. From morning to night I must fill in forms and pass them around all of Moscow—the Cheka, the Ministry of Foreign Affairs, Narkomzem [the Commissariat of Agriculture], Rabkrin [the Worker-Peasant Inspectorate], Vneshtorg [Foreign Trade] and Sovnarkom [the Council of People's Commissars].

"Now the matter is stuck with the Cheka. I am not at all convinced anything will come of it. The trouble is I am working alone, of course, nobody here has done anything for me."[3]

The bigger problem was how to pay for the seeds. The young communist state had almost no reserves of foreign currency. The old tsarist ruble was worthless and no one would change the new Red ruble, which had been subject to wild inflation. As a result, trade with Russia was frequently carried out on a barter basis, often the swap of precious metals and diamonds in exchange for goods. Officially, Britain and the United States did not accept Russian gold, but both countries were so anxious to get a piece of the emerging Russian market, especially in industrial machinery, that officials often turned a blind eye to gold transactions. In addition, a number of neutral governments, especially Sweden, "washed" Russian gold by reminting it.[4]

The other favored commodity was platinum. At the end of the Great War, platinum was in short supply and eagerly sought, especially by the British and the Americans.[5] They needed platinum for military-related supplies such as fuses and firing charges for large-caliber guns. It was also used for electrical, telephone, telegraphy, and wireless communications, and Russia had plenty of it. Nikolai Ivanovich set off for America carrying solid gold rubles and ingots of platinum.

"To cross borders with gold is such an undertaking that it still appears unbelievable to me," he wrote to Lenochka. "But I shall dare to try. Too much energy has been invested already."[6]

For all his boyish enthusiasm about this adventure, Vavilov knew its import firsthand—not only as a scientist, but from simply talking to his colleagues in Moscow. From them he learned of the terrible effects of the spreading famine. Several former colleagues from Saratov University had fled the famine, which, they reported, was particularly devastating in the Volga basin. "The harvest is worse than in 1891," he wrote to Yelena, "and from where help will come is not known." Typically, Nikolai urged her to be stoic. "It is necessary to be cheerful, calm and persistent." It was precisely when things were bad, and others gave up, that he knew she would need his inner strength. Vavilov wrote that Oleg had turned into a "nice fellow, very intelligent and better than before." Life for his mother and brother, Sergei, was "tolerable." Katya had "improved since the Saratov period, but as usual she cannot adapt herself."[7] They had grown even further apart.

"Dear and sweet Lenochka, I wish I could be with you in Petrograd which I like much better than Moscow. We shall try to organize our life as we want it. I am confident that we can do that. If only I can hold up. *I am encouraged by all who lack courage*" [italics added]. Just hang on 'til I return. Cheer up those who despair . . . help people who need help."[8]

The voyage across the Atlantic lasted a long fortnight. Vavilov did not adjust easily to the rolling of his steamer. He was miserably seasick for much of the trip, yet he continued to work, managing to write an English translation of his Law of Homologous Series for British publishers.

On his arrival in New York, the immigration official, an American version of Moscow's bureaucrats, refused to let him enter. Each passenger had to have at least three hundred U.S. dollars in cash. He only had gold and platinum, worth considerably more, but it was not acceptable. Vavilov, ever adaptable, cabled his former brother-in-law, Nikolai Makarov, who had been married to his sister Lydia and after their divorce had moved to New York. Makarov met him at the dock with the $300. It was a particularly generous act since Makarov was marrying again that very day, and he took Vavilov ashore in time to celebrate.

In the next four months, Nikolai Ivanovich completed an astonishing list of official tasks. In New York, he lectured about his new law on variations in crop plants. In Washington, he held meetings with the American Relief Administration, run by the future U.S. president Herbert Hoover. (The ARA was providing more food to the USSR than all other governments combined, at its height supplying enough for more than ten million meals a day.[9] At the U.S.

Department of Agriculture, he studied the reports of the famous American plant hunters David Fairchild and Frank Meyer, and concluded that their collecting had been too haphazard and disorganized. "The effort was not based on a single dominating principle which is necessary for such a field of research," he commented,[10] relating their activities to his own centers of origin of the main food plants around the world.

After Washington, he crisscrossed America, visiting crop research stations in eighteen states in a trip that must have presented a distressing contrast to the misery he left behind. His letters talk about meeting some of the leading biologists in America. In New York he visited the future Nobel Prize winner Thomas Hunt Morgan, who ran the legendary fruit fly lab at Columbia University where the American researchers had proved that genes were arranged on chromosomes "like beads on a string."[11]

Vavilov was constantly confronted by the material advantages of the American researchers, but if he had any thoughts about leaving Russia, as several of his colleagues would do, there is no evidence of them. Quite to the contrary, he was generously extending invitations to American researchers to join him in Petrograd. When he met Hermann Muller, the socialist genetics researcher who worked with Morgan and who would win the Nobel for his work on radiation and mutations, he asked him to come to Russia. Muller would come briefly the next year, and then for a prolonged stay in the 1930s.

Despite all the problems managing his Russian resources, Nikolai Ivanovich set up an American branch of his Petrograd Bureau of Applied Botany in a one-room office at 136 Liberty Street, in Manhattan's financial district. And, in a move that was certainly in the spirit of Lenin's New Economic Policy, he started a new Fifth Avenue public relations stockholding organization, the Association for

Promotion of American-Russian Agriculture, Inc. There were four partners, two Americans and two Russians, himself and his former brother-in-law Makarov and two Americans. The bylaws stated that the new firm would "promote the general interest of American and Russian agriculture," "study farming methods," "spread knowledge of the best means of rural betterment," and "work out a system of greater economy in handling and marketing" varieties of crops in both countries.[12] One can only imagine the reaction of the comrades in the Commissariat of Agriculture in Moscow if they had known how deeply their special envoy was engaging in capitalistic ventures. But he was also buying seeds wherever he went.

In less than a month, Nikolai Ivanovich bought 6,224 packets of seeds from twenty-six different American seed companies. Perhaps most promising, he was given native corn seed from an Indian reservation that, according to the U.S. Indian Service agent, was suitable for growing in northern Wisconsin where the season is so short other varieties would not normally mature.[13] Vavilov hoped it would be suitable for growing in Russia's northern belt. He persuaded Hoover's American Relief Administration to ship the seeds—two tons in all—to the Baltic port of Riga along with the other regular shipments of food aid. On his return to Russia in February 1922, he would be rewarded for a job well done by his election as a corresponding member of the USSR Academy of Sciences, the first step to full membership, and he was still only thirty-five.

The official Soviet government accounts show that Nikolai Ivanovich took an astonishing sixty-one boxes of seeds in his personal luggage back to Russia. These seeds would be familiar to any American gardener who spends the winter poring over seed catalogues, but they were fruits and vegetables not yet seen in Russia, a whole array of promising additions to the spare Russian diet. They came from the Santa Rosa nursery of Luther Burbank.

Luther Burbank, like Russia's Ivan Michurin, was a breeder of the old school, creating new varieties intuitively, without reference to their genetic constitution or anything as scientifically rigorous as Mendel's laws of heredity. He was a horticultural artist rather than a scientist. As a youth he proved adept at spotting a superior plant variety. Born in 1849, ten years before Darwin's *Origin of Species*, Burbank had read Darwin, but Mendel's work was rediscovered too late for him to absorb it. At any rate, when it came to theory, he remained a convinced Lamarckist, believing in the inheritance of acquired characteristics. He is now known not only for the Burbank potato, but also for a wide variety of fruits, different kinds of berries, pineapples, walnuts, and almonds. He has been credited with introducing sixty kinds of plums.

Nikolai Ivanovich had been warned that Burbank, like Michurin, did not like visitors dropping in, so he wrote ahead and received a rare invitation to the master gardener's home.

Burbank lived in a small, ivy-covered home behind a white picket fence, a classic American fantasy. The garden stretched out another two and a half acres. A painted sign at the gate provided his protection. It read, "Mr. Burbank is no less occupied than are the officials of Washington, and therefore he most humbly requests that the public not disturb him with visitations." If that did not work, Burbank's ferocious secretary was there to turn away unwanted visitors, advising them that Mr. Burbank was extremely busy. The Russian agronomist, however, was welcome and he was allowed to stay for several hours.

On his walk through Burbank's magical garden Nikolai Ivanovich marveled at huge cactuses without thorns and at the sumptuous fruits that Russians could barely imagine. He saw new varieties of asters, chrysanthemums, bright red late-blooming poppies, gladioli, blue roses, dahlias, and bush phlox. The many different fruits included plums and pears, and there were grapevines and huge walnut trees. It was like a hothouse at a botanical garden, filled with the best and most luscious specimens. In one corner there was a short-stemmed sunflower with its yellow head of seeds inclining downward, as a protection against birds. In the vegetable garden, Vavilov saw a blackberry without thorns, various types of maize and sorghum. It was "a botanical fairy tale," as he would later tell his friends in Moscow.[14]

Vavilov had heard of Burbank's extraordinary creativity in breeding new varieties and he had seen the Californian's lavish twelve-volume work, *Methods and Discoveries* (1914–15). The 1,500 color plates included one of the largest plums in the world and another sketch of a plum that dried itself while still on the tree, becoming an instant prune. Burbank was a brilliant self-publicist, tending to make outrageous claims for his products, but Vavilov was truly astonished by what he saw and spent hours inquiring about Burbank's methods. "I recall the moment when standing with my camera before Burbank amidst his flowers, I understood this living fairy tale—the force of individuality in this beautiful old man with the face of an actor, of an artist among his creations."

Vavilov was fascinated by how Burbank's artist's intuition seemed to block out his scientific curiosity.

"How does your work in selection begin?" Vavilov inquired.

"First you establish an ideal. Then you seek the ideal," replied Burbank perfunctorily, divulging no scientific secrets, or even scientific understanding.[15] Even the Carnegie Institute, which for many years funded Burbank's work, found Burbank's methods mysterious. They sent a plant breeder to write up a description of how Burbank worked. In the end, the author gave up. It was impossible to convert intuition into scientific data.

When Burbank died in 1926, Vavilov would sum up his contribution to plant selection with the generosity he was known to afford any colleague. "His intuitive creativity as an artist-breeder not infrequently ran counter to the precise genetic perspective of the contemporary establishment. Not being a theoretician in plant breeding, Burbank reached not a few mistaken conclusions in the explanation of his work. The theoretical side and the explication of the results of his work do not always find Burbank at the level of contemporary science."[16]

As Vavilov observed, it was, therefore, difficult to learn anything from Burbank—"the artist's intuition overwhelmed his research." But it was Vavilov's effort to see how nonscientists could use other ideas, unscientific ideas, that would be both his genius and, later, part of his undoing.

When he finally departed for England at the end of November, Nikolai Ivanovich left a trail of unpaid bills for seeds, plus scores of books he had bought or been given. A last letter from Dmitry Borodin, the manager of the newly established New York office of his Bureau of Applied Botany, informed Vavilov of the material still waiting to be shipped. And there was the small matter of the platinum he had left behind. One ounce fetched $80, compared with $150 during the war. Borodin wrote, "Kindly tell me if you wish me to sell and remit to you the amount realized."[17]

On the way home, Vavilov spent three days with Bateson in London and also visited his colleagues at Cambridge. About his visit with Bateson he wrote, "We've talked about everything. A whole evening we've been talking about evolution. I guess it was the most substantial discussion of my whole trip." In Cambridge, he picked up two hundred samples of Afghan, Spanish, and Portuguese wheat and saw examples of Abyssinian wheat. "A trip to Africa becomes a necessity . . . if everything I have collected reaches us, our collection of cereals will obviously be the best in the world."[18]

Soon after Vavilov's return to Petrograd, the packages started arriving from America, including books, scientific journals, and about twenty thousand seeds that he immediately put to work. Within a year he had increased the number of experimental breeding stations of Regel's old bureau from three to twenty-five. His plan was to set up stations in all the extreme geographic locations from the Arctic to the central Asian states, and from the Baltic to Siberia and the Pacific Coast. He wrote to Borodin in New York that he could "judge the state of affairs here [in Petrograd] if I tell you that I ran around for half an hour today trying to find one million rubles of Soviet money [then about 25 cents in U.S. dollars] for a tram ticket . . . and we don't have money to send you a cable, either."[19] In the same letter, Vavilov mentioned that in order to secure the right to use the Pushkin estate as his experimental farm he had to pay the Commissariat of Education a billion rubles (about 250 U.S. dollars).

This was a time when Vavilov would begin writing to his newfound colleagues abroad, becoming the great networker of international plant breeding. The letters would go all over the world—to the seed merchants Vilmorin in France, to Thomas Hunt Morgan in New York, to William Bateson in England, to a researcher in Spain. Mostly, these letters were requests for plant samples or foreign publications, and sometimes to send his own scientific papers, or to propose trips. Vilmorin in Paris was asked for his "most interesting book on beetroot." Professor Morgan was informed that Vavilov had arranged for him and William Bateson to be elected as members of the Russian Academy of Sciences. Bateson was asked to take care of Vavilov's colleague, Georgy Karpechenko, who was going to study in England. Dmitry Borodin in the New York office would be asked for "hundreds of copies of the general map of Russia" that were "badly needed"[20] and not available in Petrograd.

Vavilov was still desperately short of funds. He told his bosses in Moscow that he was unable to pay his staff, hire day-workers at the experimental farm at Pushkin, or pay for horses. There had "never been so difficult and unclear circumstances," he wrote.[21]

Even so, he managed to revive several of Regel's old publications that had been lying at the printer's since the 1917 revolution. These included works on plant breeding and bread wheats. By the end of 1922 he wrote Bateson that he was even beginning to "solve some of the problems of the origin of plants."[22]

CHAPTER 11

Afghanistan, 1924

Nikolai Ivanovich returned home more determined than ever to continue his international expeditions. Now he wanted to search for these centers of genetic diversity—nature's most prolific plant laboratories—in Afghanistan, North Africa, and Ethiopia. Afghanistan was a dangerous place, then as now, plagued with local rebellions and wary of foreigners. A Soviet prospector on official Bolshevik business was especially suspect. In his typically understated way, Nikolai Ivanovich would note that travel in Afghanistan was "rather difficult."

On June 19, 1924, Nikolai Ivanovich led the first Russian scientific expedition into Afghanistan. Although the Russian Asiatic republics had a border of some 1,500 kilometers with Afghanistan, no Russian botanist had ever collected plants in the country, despite the fact that then, as now, it is mainly agricultural. In five months, Vavilov traveled more than five thousand kilometers, almost all of it on horseback. When he searched beyond the Hindu Kush, on the border with British India, Vavilov discovered unique wheat and rye varieties unknown in Europe. He collected samples of fruits and vegetables, especially melons, and he gathered seeds of cotton. But it was wheat, the bread wheat, that turned out to be a treasure. Afghanistan proved to have the richest diversity of bread wheat in the world. In all, he would bring back seven thousand seed samples for his collection.

From the start, the expedition was plagued with problems. The Afghans were concerned that if they gave a visa to the Russians, the British would be next and then perhaps the Germans. Before they knew it, the entire country would be swarming with foreigners. To overcome the Afghan objections, the Soviet foreign ministry made Vavilov and his two research colleagues plenipotentiary diplomats; Vavilov was listed as a trade representative and his colleagues as couriers. The move made the British suspicious. They had been following Vavilov's activities closely for two months.

On April 14, 1924, diplomats at the British mission in Moscow sent a cable, stamped SECRET, to London advising of "Professor Vavilov's scientific expedition to Afghanistan."[1] They were concerned that Nikolai Ivanovich was being sent by Moscow to foment hostility on the Indian frontier where two British officers had been murdered recently. Afghanistan itself was going through a painful attempt to introduce Western values, in particular the education of women, but Islamic clerics violently opposed any change. There were constant delays while waiting for permission from local officials to proceed from one town to the next. Vavilov's first interpreter, a Russian who was picked up at the Soviet consulate in Herat, was drunk most of the time. He could not even speak Farsi, the language of most Afghan officials. "I therefore immediately had to set about perfecting my own Farsi; there was no other way out," Vavilov wrote in his notebook. "I rose early in the mornings and memorized the boring Farsi grammar."[2]

The small caravan was not allowed to travel from one city to another without a military escort, usually two mounted Afghan sepoys—the native soldiers trained by the British—who turned out to be a halfhearted, fickle bunch. They often complained that Nikolai Ivanovich was driving them too hard. On some of the more mountainous trails in the Hindu Kush, even the pack horses were afraid to negotiate the rickety river bridges and the boulder-strewn trails. Often the animals had to be led without their load and then the baggage carried separately, by hand. The sepoys rebelled: they went on strike and another time they simply disappeared for a while. At one point, they refused to accompany Vavilov because, they said, there were too many bandits on the route. The Afghans were mostly friendly, but not always. In one small town Vavilov photographed ruins that turned out to be of religious significance to the local mullahs, who rushed out and started throwing stones until the offending foreigners moved on.

Originally, Vavilov had intended to cover the northern part of Afghanistan, including the mountainous Hindu Kush, but he was plagued with bouts of malaria and one of his colleagues kept falling sick from the local food. These delays stalled the expedition and caused a constant revision of the route.

At the first destination, Herat, in northeast Afghanistan, the fields were filled with wheat, barley, millet, maize, fava beans, winter cress, castor oil, and fenugreek. There were also fields of cotton, hemp, tobacco, and opium poppies. The gardens were overflowing with apricots, pears, plums, figs, pomegranates, and peaches. Vavilov took samples of them all.

For all the wealth and variety, however, it was the presence of rye in the wheat fields of Herat that turned out to be the most scientifically important. He had seen rye growing in other wheat fields, but it was cultivated rye. This was wild rye, sometimes mingling with the wheat like a weed. This rye was evidence that Vavilov had found what he was looking for to demonstrate the origin of rye. In his report he declared optimistically, "We were at the origin of the variation of plants cultivated in Europe."

After Herat, the caravan moved slowly. It took thirteen days to travel the 540 kilometers from Mazar-e-Sharif to Kabul, staying overnight at local inns. The route was over mountains and into valleys and was "still not well blazed and only dynamite could have improved it." At last the party could see the massive peaks of the Hindu Kush as the famous range spread out like a fan from 24,000 feet.

"Suddenly horsemen came galloping up," Vavilov noted in his diary. "They halted the caravan and told us that we had to wait for their leader. It was apparent that something unpleasant was about to happen, but then we learned that their leader, the local governor, had been badly wounded in an apparent assassination attempt. He was counting on the caravan of 'Europeans' to help. At the inn there was great commotion. Night was already falling and a crowd of some hundred men had gathered outside."

The leader was brought in on a stretcher and Nikolai Ivanovich produced iodine and an English disinfectant he had bought in Mazar-e-Sharif. He boiled some water and while the governor's bodyguards looked on he applied the iodine to the wound and bandaged it, hoping this primitive first aid would work. At dawn the horsemen suddenly reappeared, but this time they only wanted to express the governor's gratitude as they handed over his personal gift of dried apricots and nuts.

Word of the European medicine men spread fast and at every inn thereafter Vavilov was met with a line of sick people seeking help. He gave them quinine or aspirin, while supplies lasted.

As they drew closer to Kabul they heard of an uprising of southern tribesmen serious enough to threaten the overthrow of Afghanistan's modernizing leader, Ammanulla Khan. Europeans were already fleeing Kabul, but Vavilov was determined to continue. "The prospect of going back to Mazar-e-Sharif when three-quarters of the mission was still not finished did not appeal to us," he noted. "We needed to reach Kabul one way or another."

Vavilov's perseverance was rewarded. The discovery of wild

rye at Herat was only the beginning. En route he found "wonderful crops of awnless wheat, alfalfa and Persian clover." (As noted
earlier, the awn is the spinelike growth that extends from the grain
sheath in cereal like wheat, barley, and oats. Those with awns are
called "bearded." The presence or absence of awns can affect yields
and is therefore of great interest to breeders.) And as they entered
the outskirts of Kabul, Vavilov spotted a high-yielding so-called
club wheat that had a particularly tough stem. He also found small-
seeded, dark-colored beans sharply different from European strains,
and also cotton varieties similar to those found in India.

Vavilov was surely breathless as he collected and catalogued so
many varieties. He noted, excitedly, that this was "a kingdom of
endemic plants. . . . It was perfectly evident that we stood at a source
of an original cultivated flora." He concluded that these plants had
developed over centuries, not in pleasant, irrigated river valleys
as once thought, but under severe conditions of stony soils and
extremes of temperature. It was a rash conclusion on the basis of
not much evidence, but it was an understandable one. He was, after
all, the first botanist to make such an observation.

By the time he reached Kabul, Nikolai Ivanovich had intended to
return home, but Afghanistan was too enticing. He was like the
archaeologist who has found the door to an ancient tomb. He could
only move forward now to see the rest of Afghanistan's botanical
treasures.

The expedition split into two, one returning to Herat via Kandahar, Vavilov and the others taking the less well-known route in the
direction of the Pamirs. It would be an even more perilous journey
than his first foray into the Pamirs in 1916.

Vavilov, a colleague, a guide, two sepoys, and three horses for
the luggage proceeded in the direction of the Salang Pass, the most
important route through the Hindu Kush, linking the capital, Kabul,

to northern Afghanistan. As they climbed they encountered last year's snow, and frozen brooks and pools. The road turned into a path and they had to walk the horses, but suddenly there was a breathtaking panorama of the ridges of the Hindu Kush. For that alone, the treacherous climb had been worth it to Vavilov. On the eighth day the caravan entered Khanabad in the valley of the Talikan River, then climbed again to Feyzabad and on to their destination, Zibak. "The intended goal had been reached, a beautiful agricultural area with irrigated fields including endemic non-ligulate wheat with simplified leaves and gigantic spring rye." They proceeded along the border with India (now Pakistan), but his colleague fell sick again and so Vavilov went on alone into Kafiristan, then back to Kabul, through Nuristan, the province that only one other European, a British Colonel Robertson, had gone before.[3]

"The climate in Zibak is severe and the settlers are poor," Vavilov wrote. "Their clothes are terrible and in spite of the cold they go bareheaded. For lack of sugar they drink tea with salt. We spent the night around a campfire and in the morning the stream was covered with ice. Horses and people sank into deep snow."

The village of Pronz at 2,880 meters seemed like a new world from the rest of Afghanistan. The women appeared to feel their own strength, were liberated and entered freely into conversation with the Afghan soldiers from Kabul. There were many children with Aryan, white faces. Vavilov was a botanist, not an anthropologist, but he noted, "The Hindu Kush appears to be a mighty barrier long ago isolating the land of the Kafirs. Their language has very different roots. I had to write a new botanical dictionary. Names differ for wheat and barley."

The path along the River Parun was "unforgettable," he wrote. "The horses' legs got stuck in cracks between the rocks. It was only possible to advance extremely slowly. There was no other choice. Hour after hour passed, one mishap following another. A horse hung down over a steep slope, its feet stuck in a fissure; the pack fell off the cliff down into the river. Sometimes we had to unload the pack and carry it by hand, pulling the horses by force down the steep slopes."

Vavilov kept to his mission even through these difficulties. "Near Vama we began to come across tiny plots with wheat, millet and sweet sorghum. Again the inhabitants were of the Aryan type. Although Muslim, the women walk around without veils and are totally liberated. The children and the men wear goatskins with the hair on the inside and no sleeves. Indeed, the first people on earth could have dressed like that."

"On October 23 we left Vama for Gursalik, but no one would agree to accompany us because there were so many bandits. By giving them five rupees in advance we managed only with difficulty to persuade four Kafirs to take us within a few miles of Gursalik, but not to follow us in. The trail was awful, suitable only for walking and for goats . . . we came to a half way broken down bridge. The first horse fell through its braided twigs. Somehow we succeeded in saving the horse and repairing the bridge with branches and stones. Then the guides went on strike, wanting to return home and even trying to give us back the rupees we had paid them. Somehow we persuaded them to stay on. The trail followed along the winding bed of the Parun river which has steep banks. We had to unload the horses here and there and carry the packs by hand part of the way. After a few more kilometers, the guides abandoned the caravan and escaped."

At the border with the British North-West Frontier, the caravan entered the restricted zone between Afghanistan and India. Vavilov knew he was not supposed to be there, especially with his quasi-diplomatic status as a Soviet "trade representative." But throughout his travels Vavilov would adopt a rather disdainful attitude to politics; science came first and politics should not be a factor in the quest to find the birthplace of wheat, or any other staple. Unless, of course, he stumbled across something interesting by accident.

Vavilov told a colleague later that when he was in the prohibited zone on the Afghan-Indian border he took "a whole album" of photographs of a British fort by approaching the fort from a little used path. His colleague was shocked. He could have gotten in trouble for spying. But Vavilov laughed off the incident. It was the wheat he was looking for that mattered, not the fort.[4]

The hospitable local governor wanted Vavilov to stay for the arrival of the English colonel from Chitral, a few miles inside India, but Vavilov saw trouble. "We had reasons to avoid such a meeting," he noted. He knew perfectly well that an English emissary would not be happy to hear that Soviet agronomists were traveling along the forbidden area of the Indian border. "We stayed overnight and hurried on towards the south, with an extra detachment of eight foot soldiers in view of the dangers on the road and as an honor guard."[5]

They passed through Jalalabad on October 27 and for the final trip to Kabul, they were assigned two cavalry soldiers in addition to the usual two sepoys.

When they returned to Kabul, Vavilov was warned of major unrest in the south, but he was determined to complete his survey. "At any cost we had to fill a large gap and collect seeds in southern and southwestern Afghanistan, in the areas of Kandahar and Farah and along the border with Iran." In the south, Vavilov collected "a myriad of medical herbs. I haven't seen so many apothecary dealers as in the south of Afghanistan, the whole clan of tabibs—drugmakers. I defined Kandahar as 'the town of apothecaries and heaps of pomegranates' . . . the like of which we could never expect to see in our own country or even beyond its borders."[6]

On December 24, 1924, six months after entering Afghanistan, Vavilov and his colleagues boarded the train for the slow trip, via Samarkand, back to Moscow. During the night when the train was being shunted onto a branch line, Vavilov was walking from the restaurant car back to his carriage. The railroad workers had forgotten to put down the bridge between the two cars and Vavilov fell, but fortunately landed on a buffer. "The more than 3,000 miles along terrible trails and mountain slopes in Nuristan and through waterless deserts turned out to be less perilous than the passage on the main railroad line!" he wrote. "I have unwillingly become a fatalist."

In Tashkent, Vavilov was asked by the Agricultural Department of the Central Asia University to give a talk about his Afghanistan expedition. Vavilov's reputation as a talented botanist and indefatigable explorer had preceded him and the students packed the hall, but they emerged disappointed. As one of them would later recall, "We had heard about this young, rising star of the scientific world, but he did not impress us. Nikolai Ivanovich spoke in a very soft voice, lethargic, with frequent pauses, obviously searching for words. He changed his pose a lot, sometimes leaning on the rostrum with his chest, sometimes clinging to it. It was only afterwards that the students learned that Vavilov had an attack of malaria and had insisted on giving his lecture despite having a fever of 40 deg C."[7]

On his return, he would tell a friend, "We cleaned out all of Afghanistan."[8]

For his work on this expedition, Vavilov would receive the Lenin Prize, the country's highest honor for services to the Soviet Union, and the Przhevalsky Gold Medal of the Geographical Society of the USSR. And he would become a member of the USSR Central Executive Committee, a small parliamentary body later replaced by the Supreme Soviet, the national legislature.

He was modest about his rewards, preferring to think about the new botanical delights in store on the next expedition. On receiving the Lenin Prize he wrote to Yelena, "In and of itself, it doesn't interest me. We are all equally [poor] proletarians. But I am touched by the attention. We'll keep trying."[9]

By the end of 1924, Vavilov's world collection had grown to almost sixty thousand accessions, including the seven thousand from Afghanistan. Each year about forty different species of cultivated plants were being planted in the experimental stations that would soon total more than one hundred. Lenin had died, Petrograd had been renamed Leningrad, and Regel's old bureau had been reorganized into the All-Union Institute of Applied Botany and New Crops. The "most valuable of all the possessions" of the old bureau to be passed on, Vavilov wrote, was its "large number of workers . . . all of us are a consolidated unit that enables our ship to move towards her goal. We set high standards for ourselves."[10]

Despite his bout of malaria, Vavilov was soon full of energy again, keeping his staff up to the mark and resuming his international networking. If his researchers failed to produce papers he insisted that they work as hard as he did. His long hours were legendary. One researcher recalled, "I was about to go to bed one night when a messenger told me that Nikolai Ivanovich wanted to see me. I got nervous as we had seen each other only three hours earlier. When I got to the Institute all the senior scientists were in Vavilov's office having tea and sandwiches. Nikolai Ivanovich was surrounded by maps, photographs, tiny bunches of dried plants, seeds on paper plates and was telling everyone about his trip to Afghanistan. The session went on until 2 A.M."[11] It was quite usual for researchers to be asked to Vavilov's home for a chat after midnight; he was always gracious and hospitable, providing cookies and a light wine, which he never touched. He rarely drank alcohol.

In letters to staff who were not pulling their weight he could be blunt. In 1924 Georgy Karpechenko was working at Bateson's John Innes Horticultural Institute outside London and had not yet written up his research. Vavilov scolded him, accusing him of laziness and declaring that it was his "sacred duty" to perform better. Soviet science was different from science in the West, he wrote. "At [the John Innes Institute] it doesn't matter whether they study flax or roaches, or whether they work on genetics or mycology. But we are not steering our huge Institute for the comfort of the employees. We understand that, with all the defects of this sort of collective, it is capable of a huge practical effort. That is the symbol of our faith."[12] One researcher was moved to another job in Moscow because of his "overt religiousness, singing in churches, and having lit-up icons in his ante-rooms."[13] In reviewing another researcher's paper Vavilov upbraided the author for giving too much credence to the inheritance of acquired characteristics, insisting that "there was no experimental data" on Lamarckism, "nothing at all."[14] But Vavilov's potato expert, Sergei Bukasov, who had been in Peru, received the harshest criticism. "Sergei Mikhailovich, we are very angry at you for not writing a word."[15]

He always had requests for plant samples, and things he couldn't buy in Russia. To a Russian researcher in Palestine he wrote, "We desperately need samples of wild wheat, *Triticum dicoecoidus*, from Hermon Mountain, near Jaffa, where it grows in abundance. You probably know that it was first found in Syria and Palestine by Aaronsohn, the director of the Jaffa agricultural station, in 1907." And he asked another researcher in China for "one or two handfuls of seeds of the following plants: barley, wheat, rye, flax, hemp, lentil, horse beans, mustard, as well as melons and cotton."

He seemed to sense that time to complete his grand scheme was running out. While he was in Afghanistan, the Commissariat of Agriculture had stopped financing Vavilov's New York office and dismissed the manager, Dmitry Borodin. Vavilov hinted that the political climate was changing. He promised that Borodin would be reinstated (which he was, to purchase more seeds). Vavilov wrote, "Eventually they are going to realize that you mustn't be removed, but it's difficult to get anything done in the fussy atmosphere existing here, even if it's obviously the right thing to do." Vavilov ended with a plea—to "show kind mercy and use what's left of my money [from the ingots of platinum] to send me three or four fountains pens. The last ones I had were worn out in Afghanistan."[16] And he needed a fresh supply for his next expedition.

CHAPTER 12

Abyssinia, 1926

Nikolai Ivanovich returned to Petrograd, now Leningrad, at the beginning of 1925. In the year since Lenin died, the "bourgeois specialists" had continued to enjoy their independence and had reason to think it might continue. The Commissariat of Foreign Affairs received Vavilov's Afghanistan report enthusiastically, and he prepared for his next expedition to southern Europe, North Africa, and Abyssinia (now Ethiopia). But the political climate was indeed changing.

For the first time, Vavilov's application to leave the country was rejected. He appealed immediately to Nikolai Gorbunov, who was still, as he had been under Lenin, the highest Soviet official for science and technology. Was his "loyalty to the Soviet power" being questioned? Vavilov asked. Surely he had already proved himself to be a strong patriot on his trips to Persia, America, and Afghanistan. He argued that these trips were the "essence" of his Institute's work. The seed material he collected had put the Soviet Union on the "level of world science, and without doubt contributes to the prestige of the USSR."[1]

What was essential now was to move forward. The Institute (as we shall now call Vavilov's All-Union Institute of Applied Botany and New Crops) needed to collect and work with the important varieties of cultivated plants to be found along the Mediterranean coast—durum wheat, oats, barley, flax, legumes, and sugar beetroot. Investing in these trips would pay off, he argued. "Paltry expenditures according to the most meager calculus, with total suppression of personal interests, will give the most valuable practical material to our seed-breeding, I have no doubt."

Nikolai Ivanovich stressed the special advantages to Soviet foreign policy of having a botanist travel in these countries and return with firsthand reports on their economic status. "During the [forthcoming] expedition, along with special tasks we can collect data of economic character, the way it was done by us during the expedition to Afghanistan."

The timing of the trip, he insisted, was crucial. It was necessary to be in those countries during the ripening season in early May. He had already spent a year trying to obtain permission from the various colonial authorities and any delay now would be "difficult and embarrassing." He was referring to Algeria, Tunisia, Morocco, and Somalia, which were French colonies, and Syria, which was a French mandate. Sudan, Egypt, and Cyprus were British, and Palestine was a British mandate. Eritrea was Italian. The diplomatic missions of these countries had been reluctant to give a visa to a Soviet professor of botany, especially one who was on official Bolshevik business. Vavilov observed in his expedition report, "One can only dream that mankind could return to the time of Marco Polo when a traveler without visas was able to cross continents and oceans to the destination planned by him and be received everywhere as a welcome guest."[2]

In Moscow, the letter to Gorbunov had the desired effect: permission was granted. By early summer, preparations were complete. His baggage included the best available maps from Stanford's, the London cartographers, a selection of aneroid barometers (for reading height above sea level), several rather large botanical reference books, and his precious camera. But he still had to get visas.

His journey began in London where Bateson, who had died of a heart attack, had left instructions for his staff to assist Vavilov. With their help he managed to get visas for the British territories of Palestine and Cyprus, but not for Egypt and Sudan. When his visa applications to the French Ministry of Foreign Affairs were ignored, he appealed to Madame de Vilmorin, of the Vilmorin and Andrier plant breeding company. He had met her briefly during his stop in Paris on the way back from London in 1914 and she had taken a liking to the young Russian. Madame de Vilmorin did not consult a friendly bureaucrat; she went directly to the French president, Raymond Poincaré, and to the prime minister, Aristide Briand. The matter was complicated by a rebellion in Morocco and an uprising of the Druze in Syria. The French view was understandable—they were wary of a Soviet professor peddling Bolshevik propaganda in their colonies at such a perilous time. Even French citizens were barred from traveling in Morocco and Syria. But Madame de Vilmorin worked her magic. Vavilov was given access to Lebanon and Syria, and the French colonies along the North Africa coast—Tunisia, Algeria, and Morocco.

By mid-June Vavilov was on a boat from Marseilles to Algiers. Over the year that this expedition would eventually take, he sent thousands of rare seeds to his seed bank in Leningrad. Despite his insistence to officials that this trip would be carried out "with total suppression of personal interests," Nikolai Ivanovich had planned a very personal rendezvous. Katya had finally agreed to a divorce and would stay living with Oleg and Nikolai Ivanovich's mother, Alexandria Mikhailovna, in Moscow. He could now admit to his secret love for Yelena and they could live together. They would never get married, and strangely, in the great outpouring of recollections from colleagues about Nikolai Ivanovich's life that followed his rehabilitation, there is almost no mention of Yelena Barulina. Yet it was a great romance; they loved each other deeply, no doubt. And perhaps, the romance was enhanced by the secrecy, by the hardships they endured during those first years together. In all the drudgery, it must have been uplifting to have a secret affair. Before leaving Leningrad, he arranged for Yelena to spend several months in Germany as a researcher. This was a way of getting her out of the Soviet Union and enabling him to meet her in Italy on his return from North Africa.

For two months, Vavilov traveled in the French-controlled territories along the North Africa coast. It was a profitable time. He met French plant breeders in Algiers and Tunis and went up into the Atlas Mountains in Morocco.

"So this is Africa," he wrote in his diary on arrival in Algiers. "But there is very little of the real Africa left here." He was referring to the foreign plants introduced by Louis Trabut, the famous French botanist, who had planted "everything of value within the plant kingdom from tropical and sub-tropical climates—Peruvian philodendrum, Australian eucalyptus, acacias, citrus trees from south-eastern Asia, Mexican cacti and endless vineyards stretching in all directions."

Trabut, then seventy-three, was a man with much the same grand vision for Algeria that Vavilov had for the Soviet Union. Trabut welcomed his Russian colleague, showed off his herbarium, and the two of them worked out an itinerary. Vavilov was impressed by Trabut's modest existence, especially for a capitalist. "It showed that even in the rich countries the work of scientists is to a great extent unselfish and is in no way remunerated according to the value of the results achieved."

As always, Nikolai Ivanovich was impatient to be out in the fields. Trabut warned him that only "mad dogs and Englishmen" go to the desert oases in July, but, "to hesitate was out of the question,"

Vavilov retorted. With an Arab guide, Nikolai Ivanovich drove deep into the desert. The trip was hot, difficult, and, at first, disappointing. Although he saw plenty of date palms, their quality was not as good as those in California.

In the markets of the coastal towns, however, Vavilov found the biggest vegetables he had ever seen, including onions weighing four pounds—the result of centuries of careful cultivation.

From Algeria he took a bus to Morocco where he hired guides and horses to go into the Atlas Mountains. There he discovered a peculiar variety of hard, or durum, wheat as well as rye, hemp, pea, and cowpea. Morocco's original plant life so fascinated Vavilov that he overstayed his two-week visa and instead of going overland to his next stop, Tunisia, he had to cadge a seat on a French military plane. Soon after takeoff the plane developed engine trouble. "Below the aircraft was a lifeless desert. The pilot performed some violent maneuvers, diving and turning to try and re-start the engine. In the cabin a French officer and I were thrown against each other and the walls and we were finally delivered to Oran half-conscious. There I had only a few hours to recover before catching my train [to Tunisia]."

In Tunisia he teamed up with the French botanist Professor Félicien Boeuf, "who had planned a most interesting itinerary throughout the country, including all the main agricultural areas." Again he found durum wheat and six-grained barley. At the end of the trip, Vavilov concluded that the main cereal crops of coastal North Africa had originated from the Near and Middle East—his next destination.

In Beirut, the French authorities had trouble understanding how a Soviet citizen had a visa when French citizens were being denied access to the country. All the mountainous land south and southeast of Beirut—exactly where he planned to go—was under martial law because of the Druze uprising. The military commander strongly advised Vavilov to travel with a white handkerchief on a pole as a sign of his peaceful intentions, although there is no evidence that he did so.

For the next two months, Vavilov toured Syria, Palestine, and Trans-Jordan. In Mesopotamia [now Iraq] and the valley of the Euphrates the harvest was at its peak and he found drought-resistant varieties of wheat that would later be of special interest to Ukraine.

Vavilov had an explorer's curiosity and would often comment on other aspects of life, besides botany—such as archaeology, culture, and, on rare occasions, politics. He was appalled at the effects of French control, finding Syria in a period of "profound decline." He noted the lack of effort by French officials. "In all of large Syria, which exceeds even the size of France itself, I found only one agronomist, a Mr. Ashar, who at the same time was consultant to the Syrian government on economic problems. . . . As a mandated territory Syria is a typical example of the political economic absurdity that still dominates the earth."

In British Palestine where at that time only 20 percent of the land was settled by Jews, he found the behavior of the British authorities "highly peculiar . . . they defend mainly the rights of the Jews, but at the same time they prevent them from penetrating into Trans-Jordan, which is also a British mandate." Jewish botanists were not allowed to travel with him on a field trip to Trans-Jordan.

He swam in the Dead Sea, saw Jericho, admired Jaffa orange groves and date palms in Gaza, but he concluded, "It is impossible to escape the detrimental effects of the national disunity and the discord that seem to be what is primarily being cultivated in this country."

At the end of two months in the Middle East, Vavilov took a boat through the Suez Canal down the Red Sea to Somalia, his entry point into Abyssinia. On December 27, 1926, Vavilov boarded the train to Addis Ababa. As the two locomotives pulled the train slowly out of the Somalian savannah onto the Abyssinian plateau, Nikolai Ivanovich began to see fields of wheat passing his window. Leaving the train without a permit was not allowed, but Vavilov could not resist. When his train stopped for the night, he left it and went plant hunting for two days, the authorities apparently turning a blind eye. He stuffed his specimen bags with all he could find, some he had never seen anywhere else, including types with violet-colored grains. Who knew what genetic treasures they might hold? He also found the native Abyssinian teff, a kind of small millet that produces a "first class flour used for flat cakes."

When he finally arrived in Addis Ababa, he was told that before traveling further into the interior he had to obtain an official permit. With the help of a friend of Madame de Vilmorin he was soon face-to-face with the Regent Ras Tafari, later to become Emperor Haile Selassie. The two men got along splendidly. As Vavilov was outfit-ting his caravan he was invited for an evening alone with Ras Tafari. "He questioned me about my country. He was especially interested in the revolution and the fate of the imperial court," Vavilov wrote. "I told him briefly what was known about it. It was difficult to imag-ine a more attentive listener. The ruler of Ethiopia listened as though it were a fascinating fairy tale."

As a result, Vavilov received his entry papers in ten days—a zoological expedition from the Chicago Field Museum had to wait five weeks. Hearing of Vavilov's good fortune, other envoys seeking permission to trade in Abyssinia sought out Vavilov, and he became the toast of Addis. The Italian envoy, on learning that he intended to go to Eritrea after Abyssinia, lent him a servant. This allowed Vavilov to refresh his Italian—"especially important in view of my contemplated expedition to the Italian colony of Eritrea, from where I planned to return to Italy." He did not mention that it would also be useful for his planned holiday in Italy with Yelena.

On February 17, Nikolai Ivanovich wrote to Viktor Pisarev, whom he had put in charge of the Leningrad Institute while he was away, "The Ruler of Ethiopia, Ras Tafari, whose portrait I enclose, gave permission to enter the depths of the country. Today that caravan (11 mules, 12 men, 7 guns, 2 spears, 2 revolvers) sets out toward the upper reaches of the Nile. . . . If crocodiles don't eat us as we cross the Nile, we hope to be in Asmera [Eritrea] by early April."[3]

The official papers stated that Vavilov was an honored guest of Ethiopia, and that local rulers were ordered to render him full assistance, to provide him with ammunition and provisions, and to allow him to cross their borders. Vavilov planned to be on the road in the interior of Abyssinia and Eritrea for at least three months. He tried to give his men sandals, purchased in the market, but they preferred to go barefoot. According to the local custom, every traveler had to agree to feed his men, look after their health, and three times a month give them a de-worming medicine. In case of death, he agreed to bury them in a fitting manner. Worried about discipline, Vavilov asked for advice and was told to take shackles, just in case.

On February 17, 1927, Vavilov's caravan set out on the road to Ankober in the major agricultural region. The progress was slow, on average between thirty and forty kilometers per day. At night, they camped in tents and Vavilov would put the collected material in order and write his diary. The high plateau is cut by deep canyons, which the caravan negotiated with great difficulty. At the bottom of one was the Blue Nile, teeming with crocodiles.

"We sent out guards, who began to fire into the water to drive them away. It was as if a war had broken out. Some of the crocodiles floated to the surface, belly up, as the caravan slowly made its way to the opposite shore. Frequently, as a warning, more shots were fired into the water. No one was harmed, not even the horses."

The caravan mostly stopped at army posts where the officers

were lavish with their hospitality. "Long ceremonies begin. The host is obliged not only to feed the caravan, but of course also to make it drunk. At one garrison, the commander wanted me to stay a couple of days, showing me a captive lioness and suggesting a hunt. All this was wonderful, but we had to hurry." On another night, one of the crew got drunk and became violent. Reluctantly, Vavilov shackled the man. "Toward morning, he came to himself again and all ended comparatively happily."

For all the crocodiles and the drunkenness, Vavilov himself made what he called a "first-class discovery," a peculiar awnless hard wheat, previously unknown. "For decades, plant breeders from different countries have tried to produce an awnless hard wheat by crossing ordinary awned hard wheat with soft awnless wheat." In Abyssinia, nature herself had done what agronomists could not, and Vavilov collected thousands of ears. "From both a practical and a theoretical point of view, this was, indeed, the most interesting find during all the time spent traveling in Abyssinia," he noted.

On the final trek into Eritrea, Vavilov ran into a group of armed locals who blocked their way. The caravan had been looking for a place to stay the night, but the look of these men "did not inspire great confidence," as he put it. "A consultation took place in our tent. There was without doubt a danger. In the best case, we would only lose our mules. The council decided to present the leader of the gang with two bottles of brandy, the last ones remaining from our supply for special occasions. In case this did not have the desired result, we would pay them off with silver coins." The brandy was received and quickly finished, the bandits soon dozing off. They were still sound asleep when Vavilov and the caravan quietly slipped away before dawn.

When the caravan finally reached the capital, Asmera, the last stop, Vavilov surrendered his weapons, as they were "no longer needed," and then mailed his treasured seeds. He wrote to Pisarev, whom he was now addressing playfully as "Your Excellency."

"It is my honor to report that three days ago I completed sending materials of the expedition from Abyssinia. Four days and nights I wrote non-stop. My hands got numb from signing eight hundred and thirty forms for the customs officer, seven forms per parcel. I sent fifty-nine parcels and before that sixty-one parcels from Addis-Ababa, from Djbouti and Dire-Daya, one hundred and twenty parcels in all from east Africa. The governor of Asmera arranged a special dinner for the Soviet professor, who was forced to wear a frock coat and tails."[4]

Then he added, "I'm off to Italy around April 16." He had planned to meet Yelena in the first week in May.

In summing up the trip, Vavilov classed the region as a "deservedly independent center, including special crops—teff, bananas, mustard, violet-grained wheat, and the 'absolutely original' species of awnless wheat, testified to this assessment." But it was not one of his rich centers, he concluded. There was no wild wheat, or wild barley or beans. These had come from elsewhere. For all his efforts, he reported, "The variety of the crops is comparatively poor. There were none of the fruit trees so characteristic of southwestern Asia, Mediterranean, eastern Asiatic and Indian centers." The vegetables were also "miserable."

Before returning home he went to Greece, Crete, Cyprus, and Spain where he was followed by two police agents of the dictatorship of Miguel Primo de Rivera. They became so exhausted trying to keep up that they finally gave in, approached him directly, and asked for his itinerary. By mutual agreement the followers and the followed would catch up with each other at night. As he passed along the northern Mediterranean, Vavilov wrote about his visits to museums, art galleries, and ancient monuments in addition to his botanical observations. Although he almost always managed to include scholarly references to ancient forms of agriculture, much of his diary has the suspicious tone of a travelogue. Once more one has to wonder what his masters in the Commissariat of Agriculture in Moscow made of it all.[5]

Finally, Vavilov managed to take time off with Yelena in Italy— the "honeymoon" for their common-law marriage. She was waiting for him at a hotel in Rome where he attended an international conference on wheat, and would be awarded a gold medal for his contribution. After the conference, they traveled together to Messina, Palermo, Florence, Milan, and Venice. It would be an interlude without any interruptions, apparently, and the only record of their being together is a photograph taken outside a villa somewhere in Italy. He is dressed in his customary dark gray suit, wearing his fedora, and has his camera strap across his chest. She is wearing a dark skirt, a jacket, and a cloche hat. She is clutching a bouquet of flowers.

Within a year Vavilov increased the number of experimental stations across the Soviet Union to 115. They now stretched from Murmansk in the north to Turkmenistan in the south, from Kaunas, Lithuania, in the west, to Vladivostok in the east. At the Pushkin experimental center outside Leningrad, he had set up the Institute's Department of Genetics and invited a brilliant young geneticist, Georgy Karpechenko, to head the department. He wrote to Karpechenko, "It is absolutely clear to us that the amount of work associated with genetic issues that we are now facing is tremendous. In practice, this work has already been launched in wheat, in barley, in oats and in the Cruciferae [the cabbage family]. The work in horticultural crops, melon and strawberry will come next."[6]

CHAPTER 13

The Barefoot Scientist

Nikolai Ivanovich returned home with stories of his daring exploits, still convinced that his worldwide seed collection could overcome the scourge of Russian famines. But this time there were no accolades, no explorer's medal. The harvest was once again inadequate. A few skeptical researchers at the Institute, in Leningrad, and elsewhere in the Soviet field stations, had started to question Vavilov's grand plan. At the same time, the party spotlight was turned on a new Soviet phenomenon, the "barefoot scientists."

Until the summer of 1927, Nikolai Ivanovich had not paid any attention to Trofim Denisovich Lysenko, and there was no reason why he should have. Lysenko was a minor functionary in a plant breeding station in Gandzha (also Gyandzha) in Azerbaijan. His job was to plant green peas to see if they would grow through the mild winters, providing forage for cattle and a "green" manure, a rich natural compost for spring planting.

Then suddenly, on August 7, 1927, *Pravda*, the party's official daily newspaper, published a glowing article about Lysenko. The

130

young man, now twenty-nine, was extremely lucky on his first try. His crop of peas survived the winter, producing a lush green carpet of plants, good animal feed, good manure—and excellent political fodder for *Pravda*. Here was a young "barefoot scientist," a practical researcher who had not attended a university, who did not toil in a laboratory away from the land studying the "hairy legs of flies," as the *Pravda* correspondent put it, but "went to the root of things."

The correspondent wrote lyrically about Lysenko's triumph in turning "the barren fields of the Transcaucasus" green, so that "cattle will not perish from poor feeding, and the peasant Turk will live through the winter without trembling for tomorrow." Lysenko, the paper declared, had opened up a "completely new kind of life for the local population." The extolling of the young Lysenko was not simply a nicely-crafted feature from down on the farm. It was a swipe at ivory tower academics and an indirect attack on Vavilov, his world seed collection, and the researchers in his Leningrad Institute.

The message was clear. Here was a shining example of what socialist agriculture could look like—with such young, dedicated, practical peasant technicians like Lysenko, out there in the fields, unencumbered by academic theory, making a difference today, not in ten years' time. This was how the great Soviet state could move forward. Even Lysenko's appearance fit the portrait of the new Soviet practical farmer. He was skinny, unsmiling with prominent cheekbones, wearing a simple cloth cap and smoking a cigarette. "If one is to judge a man by first impression, Lysenko gives one the feeling of a toothache; God give him health he has a dejected mien," wrote the journalist. "Stingy of words and insignificant of face is he; all one remembers is his look creeping along the earth as if, at very least, he was ready to do someone in."

For all the praise the correspondent heaped on the young man's achievement, he also sensed the dark side of Lysenko's character. He was joyless. "Only once did this barefoot scientist let a smile pass, and that was at the mention of Poltava cherry dumplings with sugar and sour cream." (Poltava is a town in Ukraine about 450 miles south of Moscow.) But no one could have told at this stage that this simple peasant propagator of peas would soon become a monster, a driven, conniving, opportunistic henchman, willing to falsify his scientific experiments to satisfy his political masters, viciously attacking his colleagues, branding them enemies of the people, watching them publicly harassed, imprisoned, and even executed, and playing the lead villain in the Soviet Union's attack on biological science.

In singling out Lysenko, the *Pravda* correspondent was only doing what his editors wanted and Lysenko was indeed an excellent choice for party tub thumping. He was not entirely unschooled in horticulture, like Michurin. He was born in 1898 into a peasant farming family in Karlovka, in Poltava province, where it was customary for peasant children to work in the fields rather than go to school. But Lysenko was bright and more industrious than his fellow field hands, and his father, Denis, allowed him to spend two years at the village school. There, he learned to read and write at the age of thirteen.[1] A promising student with a good memory, he went on to study at the elementary horticultural school, which would have been an easy step to becoming a gardener on a large estate. But Lysenko was too ambitious to labor anonymously for the dying Russian nobility, and he took the entrance exam for the Vocational School of Agriculture and Horticulture in Uman, a century-old establishment founded on a nobleman's estate and one of the most distinguished schools of its kind in Russia. Lysenko failed the entrance exam at the first attempt in 1916, getting poor marks in religious studies. The next year, he tried again and passed.

After graduation he gained practical experience on a government experimental station and actually published two short papers, one on tomato breeding, which was to become a favorite pursuit of his, and the other, with a co-author, on new methods of breeding sugar beet by grafting buds, rather than the traditional way of planting shoots. The science of the papers was competent, if ordinary, and the writing clear.[2]

Pravda proclaimed the green peas a sensational agronomic idea—winter planting—that was gathering force among farmers. Indeed, after the publication of the *Pravda* article several visitors arrived at Gandzha,[3] and Lysenko took full advantage of the attention, managing to elevate his simple idea into a scientific theory about the role of temperature in the development of a plant.

It was essential that the peas he was using should ripen quickly in order to mature before the frosts and he had used an early-ripening variety from Ukraine. But in the milder climate of Azerbaijan he was mystified to find that these early-ripening Ukrainian peas were, in fact, slow to ripen. Other varieties that were late ripening in Ukraine had ripened early in Azerbaijan.

Lysenko proposed that the most important environmental factor in the development of a plant was temperature and his efforts to find the right variety for Azerbaijan, studying the influence of planting dates and therefore of temperature, led him within a year to his "discovery" of a process he would call *iarovizatsia* from the Ukrainian *iar*, which means spring. In English it translates as "vernalization" using the word "vernal," which means of or relating to spring.

What Lysenko had really done was a classic case of rebranding. The word was new but the process was hardly a scientific discovery. As early as 1858, Russians had reported the phenomenon. A German scientist named Gustav Gassner had examined the phenomenon in his laboratory toward the end of the First World War.[4] Americans, experimenting in the field with winter varieties of rye or wheat, established that winter varieties could produce a higher yield than spring wheat. The problem was that these plants could be lost if the winter was too harsh; if ice formed on the fields they could "suffocate" beneath the frozen crust. In the winter of 1927–28, 90 percent of the winter wheat would be lost in some regions of Ukraine—the worst destruction in nearly forty years.[5]

Lysenko then began to examine the effects of heat on winter and spring varieties. Spring grains needed a higher total amount, or sum, of heat during the early stages of germination than winter varieties, and he argued that by altering the temperature in those early phases a spring variety might be turned into a winter variety and vice versa. But all this was no more than interesting conjecture. His study suggested no practical applications because the data was insufficient. The best one could say was that he aspired to accepted methods of scientific research.[6]

In 1928, he would explain vernalization in a monograph entitled "The Influence of Temperature on the Length of the Development Period of Plants" and, with the continued help of *Pravda*, nurtured his erroneous reputation as the discoverer of the process of vernalization. In the monograph he suggested that a certain amount of heat is needed to get plants to pass through each stage of their development and a good way of expressing it in experimental form was as a sum of heat.

Nikolai Vavilov knew from his studies at the Petrovka about vernalization, although it did not have that name. He had also studied the sum of heat in plants needed in plant development. The breeding of early-ripening varieties was high on the agenda of Vavilov's Leningrad Institute as he tried to find crops that would survive in the northern regions. In 1922, he had created a department of plant physiology—studying the effects of both heat and light on plant growth. The department was headed by one of his senior staff members, Nikolai Maksimov, who was an experienced agronomist, seven years older than Vavilov. Maksimov's lab was located at the Pushkin experimental station outside Leningrad. An even more senior specialist in plant physiology, Vladimir Liubimenko, was especially interested in plants' reaction to light and worked at the Leningrad botanical garden. The idea of heat sums was also widely accepted at the time. Another of Vavilov's co-workers, Gavril Zaitsev, a cotton specialist, had studied the effect of temperature on cotton development and also addressed the issue in a book, published in 1926.

A colleague of Vavilov's, Professor Nikolai Tulaikov, visited Lysenko after the *Pravda* article and they had discussed Zaitsev's work on cotton. Lysenko had referred to it four times in his 1928 monograph, even comparing his conclusion to Zaitsev's and remarking how similar Zaitsev's conclusions were to his own. Even so, Lysenko was making a perfectly respectable contribution to an active field of research. He could have been, and was, forgiven at the time for being a brash young self-promoter, using *Pravda*, and Zaitsev's work, as a leg up into a scientific world without the formal training usually required to join. But for him it was the beginning of a lifetime of opportunism, taking advantage of the politically correct science of the day, not to form his ideas—he did that on his own—but to push him up through the ranks to the highest levels of the administration of Soviet agriculture.

The Soviet system had two words for recognizing peasants and workers who showed promise and promoting them as fast as possible. One was *udarniki*, or shock workers, those who worked hard and rose above their peers. The other was the *vydvizhentsy*, the "pushed ups" or "promotees," crammed hastily, and incompletely, through what should have been a rigorous academic training with the aim of replacing the politically unreliable bourgeois specialists as quickly as possible. Although he was never a party member, in his way Lysenko was both a *udarnik*—his hard work in the fields was recognized by *Pravda*—and a *vydvizhenets*—he was pushed up by the local comrades beyond his capabilities. There is no particular evidence that he imagined himself in this role, or that he sought it, as he planted his winter peas in Azerbaijan. But when the opportunity arose he recognized it immediately, his ambition was boundless, and, like his political masters, he did not care what means were required to achieve his ends. At first sight, the *Pravda* correspondent had defined his character correctly.

•   •   •

The publicity given to Lysenko from the official party organ posed a
problem for Vavilov: how to deal with this person who seemed to be
getting a lot of credit for other people's work? More generally, how
to deal with all the new intake of peasants and workers being pushed
up from below? As a purist, Nikolai Ivanovich never rejected an
idea until it was proved to be wrong. As a researcher he was always
on the lookout for new methods to test the seeds of his world collection. Perhaps Lysensko's naked self-promotion should have given
him pause, but he was less concerned about who had discovered
what and when than he was in making sure that he gathered the best
and latest research at his Institute.

He applied this open-mindedness to all his scientific work.
Although he believed Mendelian genetics should be the basis of
plant breeding, he would not dismiss contributions from the "artists"
and the "old gardeners" like Russia's Ivan Michurin and America's
Luther Burbank. He was also ready to encourage the new generation of Soviet breeders like Lysenko who were, to his mind, useful
but out of their depth. They were poorly educated. They did not
understand genetics. They were unable to speak the international
languages of science like English and German, and thus they relied
on other, simpler theories. He admired their ability to select a useful
plant from what they could observe—such as a plant's reactions to
temperature and light—rather than on theoretical calculations of
the behavior of its hidden genes. Researchers like Lysenko were to
be encouraged, he argued. This open-mindedness was part of his
philosophy.

Often Vavilov would surprise colleagues he had just met by
demanding to know, "What is your philosophy?" By which he meant
philosophy of science. He believed that science had to be used as a
progressive social force. Science had to be constantly verified and
vigorously applied in the service of humankind. How his philoso-

phy would jibe with Marxist theory, how it would deal with poorly educated pushed-ups, how it would cope with the demands of an increasingly inward-looking society—these were the problems that Vavilov would face as the revolution entered its second decade.

Outside the Soviet state, young scientists in Europe and America were watching closely to see how science would develop in the new socialist experiment, whether the new administrators like Vavilov would be given the freedom and resources the Bolshevik rhetoric promised, and how they would progress.[7] And Vavilov, the consummate internationalist, was watching them, following their debates and their publications.

One book in particular had caught Vavilov's eye. It was by the British science writer and editor Richard Gregory, who for many years was the editor of the journal *Nature*. Gregory was not sympathetic to Marxism, but he abhorred what he viewed as man's "shameful" use of science in the service of industrial capitalism—the dark Satanic mills—and in new, terrifying weapons of war. In 1916 he published a book, *Discovery, or the Spirit and Service of Science*, in which he discussed the role of scientists in uncovering the laws of nature. Nikolai Ivanovich arranged for the book to be translated into Russian—by his wife, Katya—and it appeared in Moscow in 1923. There are many passages in the book that coincide with Vavilov's own philosophy of science, as it was developing during these crucial years.

Despite his strict upbringing in the Orthodox Church, Vavilov had been an atheist from an early age. If he worshipped anything, it was science. Gregory had advice for those who would become disciples of science. "Go into the fields of Nature and labor. . . . No amount of learning from books or of listening to the words of authority can be substituted for the spade-work of investigation. New treasures can never be secured from Nature without effort;

'tribulation, not undisturbed progress, gives life and soul, and leads to success, when success can be reached, in the struggle for natural knowledge.'" Gregory railed against the rise of "ready-made opinions" that destroyed independent thought and critical inquiry. Of Mendel's rules of inheritance he warned more work was needed before Mendel could be accepted as "more than a partial interpretation of the operations of animate Nature." The laws themselves were "not so simple, and they cannot be applied to life in general until much more knowledge has been obtained of essential dominant or recessive characters in organisms and their physiological meaning."[8]

Nikolai Ivanovich agreed—about Mendel specifically and in general about the principle: research had to be constantly subjected to criticism and review. In was in this spirit that in the fall of 1927 he sent a specialist to Gandzha to take a closer look at Lysenko's work.[9] The report was not encouraging.

Lysenko received the envoy enthusiastically, even giving up his bed and sleeping on the floor. The envoy reported to Vavilov that Lysenko was an "experimenter who was fearless and undoubtedly talented, but he was also an uneducated and extremely egotistical person, deeming himself to be a new Messiah of biological science."[10] This trait was also noticed by Lysenko's collaborators, who said that Lysenko considered established academic courses to be "harmful nonsense," and told his colleagues that success in their work depended on how soon they could forget what they had learned in horticultural school and "liberate ourselves from this narcotic."[11]

While he disagreed, Vavilov was attracted by Lysenko's youthful enthusiasm and his apparent dedication to his science; he saw in him the same obsession he himself had with plants. Vavilov would have warmed to this description of Lysenko by one of the barefoot scientist's co-workers:

"He was tall and thin and always covered in mud. He threw his cap on in one quick movement and it would sit askew on his head. He paid no attention to himself or to his appearance. Goodness knows when he slept—when we came out to work he was already in the fields, and when we returned he was still there. He would spend all his time fussing around with his plants. He knew them and understood them perfectly and seemed even to be able to talk to them, delving right inside them. His plants would "want" things, "demand" other things, would "like" this and "be upset by" something else.[12]

Vavilov was prepared to overlook Lysenko's lack of education as a deficiency that could be overcome. So he proposed that Lysenko should visit the experimental station at Pushkin, thus ensuring his talents and hard work were put to good use and his experiments subjected to proper criticism. But the proposal met strong opposition from the Institute's specialist in vernalization, Nikolai Maksimov, who complained that Lysenko's work was messy, his methods unscientific, his theories only superficially tested, and his knowledge of international scientific literature nonexistent. The last point was known to carry great weight with Nikolai Ivanovich, who used to tell his staff, "A scientist who doesn't know a foreign language will never be successful . . . memorize twenty words each day before breakfast."[13] Even so, Vavilov set aside his doubts and pushed hard for Lysenko to be admitted to the Institute. The protests from Maksimov and others were too strong, however, and Vavilov relented. Lysenko remained outside the circle—where he would become even more of an irritant with every season.

It was at this time, in the autumn of 1927, that Nikolai Ivanovich experienced his first real taste of the coming clashes between the bourgeois specialists and the new scientific cadres, graduates from the party-run academies that Lysenko would come to represent. There was resistance to Vavilov's grand plan for a world seed collection, especially from outside Leningrad, from the breeding stations in the far-flung republics. The sugar beet breeders in Ukraine, where Lysenko worked, were questioning Vavilov's project. In Leningrad at the Institute there were rumblings from the new cadres of graduates about his long absences on expeditions; complaints that he was spending too much time on a purely academic pursuit, pinpointing the centers of origin of these crops and filling in the taxonomic gaps in his Law of Homologous Series. Some of them wanted him to spend more time searching for "miracle plants," to make perfumes and rubber that could create riches. While he was away in Africa, the presidium of the Institute, the ruling body composed of about twenty scientists, had issued an official reprimand in the press, accusing Vavilov of an inability, or perhaps an unwillingness, to organize such work. Vavilov wrote numerous letters defending himself and his Institute, but they were never published.

The criticism found him unusually vulnerable and rudderless. Perhaps it was the moment of real happiness he had found with Yelena. Perhaps he looked into the future and instead of a socialist haven for science, he saw nothing but trouble. Perhaps it was the sudden loss of his father.

Nikolai Ivanovich had met his exiled father at least once, in Berlin, in 1927, and had also tried to find a job for him back home. In 1922, he had written to his agronomist colleague Nikolai Tulaikov, asking if he could arrange a visa. Tulaikov was on a trip to America at the time and would be returning through London. "I have big favor to ask," Vavilov began. "My father, Ivan Ilyich Vavilov, who emigrated in 1918 is in Bulgaria at the present time and dreams about coming home. He has had difficulty getting a job without any foreign languages and the best thing is for him to come back. He is a man of no party affiliation, and I totally vouch for his loyalty. He had wide industrial experience and could still be useful. If you are in London, my most humble plea is for you to help him get a Russian entry visa."[14]

That fall, Ivan Ilyich had come home, exhausted, sick, and poor. His business ventures in Bulgaria had failed. Within a few weeks he died and was buried in the Aleksander Nevsky cemetery in Leningrad. The only record of this sad event comes from Nikolai's nephew Alexander Ipatiev, son of Alexandra, Nikolai's older sister, who recalled that all the adults went to Leningrad—Alexandra, Sergei Ivanovich, and babushka (Alexandra Mikhailovna). Nikolai was already there. "They returned in about ten days with photographs of granddad and with his belongings. I inherited his suit, a gray hat and a tie. It was my first European suit and I was rather proud of it, even though it hung on me like a sack. I also became the owner of granddad's enormous suitcase which, according to family lore, he used to take on his business trips to Nijny Novgorod Fair before the revolution."[15] If Nikolai Ivanovich ever recorded his own thoughts about losing his father, they did not survive.

In reply to his critics, and rather than confront his opponents any further, Nikolai Ivanovich decided to resign as director of the Institute and confine himself to scientific work. In late November he told Nikolai Gorbunov of his decision. Whether his intention was to silence his critics by obtaining official support for him to stay on, or whether he really wanted to leave, is not clear.

Nikolai Ivanovich explained his decision this way: "Apparently, a number of representatives of the Republics see themselves as prosecutors in a trial over the Institute. In my vision a truly sensitive scientific-research institute must move several years ahead of the times, not drag its tail or get out of step to march in unison with the fashion of the day: luffa sponges today, perfume plants tomorrow, caoutchouc [rubber-bearing plants] the day after tomorrow. . . . I think that primarily we must look for new varieties of the major existing crops, already cultivated plants."

The administration of the Institute was "a very bulky one in today's difficult environment," and meant he did not have time for his scientific research. "I never sought administrative promotion and I see myself instead in the laboratory and in the field as a scientific director. . . . I have an enormous backlog, nearly ten books that I need to complete." He said he attributed "far greater importance" to the scientific side than to lengthy sessions of the Commissariat of Agriculture and "other hierarchal levels." He asked to be relieved of the duties of director from the beginning of 1928 and to stay on "in a modest role of a scientific specialist, head of the department of field cultures at the most, but altogether without any managerial role." If Gorbunov replied, the letter did not survive, but Vavilov's plea was rejected and he stayed on as director.[16]

Within a year, Vavilov was caught at the center of a confrontation between the bourgeois specialists and the new cadres of barefoot scientists like Lysenko, that, in the end, would destroy him personally and destroy the study of genetics in the Soviet Union.

CHAPTER 14

The Great Break

Stalin declared 1929 the year of the "Great Break with the Past." His forced introduction of socialism included the collectivization of agriculture and a cultural revolution that would expand higher education to the masses, purging the old guard in the Academy of Sciences and promoting workers and peasants into administrative positions.

After the government had rejected his resignation, Nikolai Ivanovich threw himself back to work with renewed vigor. It was a measure of his genuine confidence in Soviet science and also of his patriotism. In Vavilov's mind the two would often go side by side. He would risk his life on expeditions to bring back precious plants for his Leningrad collection, and at the same time win over to the Soviet cause foreign politicians, government officials, diplomats, and other scientists with his charm and his enthusiasm. Moscow could not have had a more effective ambassador during these difficult years.

At home, Vavilov was no less active in selling Soviet science. In January 1929, he organized a gala event in Leningrad, the First Soviet Congress of Genetics, Selection, and Plant and Animal

Breeding. It was designed to mark the progress of Soviet biologists in building the new Soviet state and to launch the agricultural goals of Stalin's first Five Year Plan—a 35 percent increase in average grain yields per hectare.

Outside the hall a banner declared "Spread the Achievements of Science Among the Masses." Nearly three hundred papers were given by Soviet scientists and a handful of distinguished foreign plant breeders and geneticists, including Richard Goldschmidt and Erwin Baur from Germany, and Harry Federley from Finland. Party leaders came to give the congress their blessing: Sergei Kirov, one of Stalin's close aides and the party chief in Leningrad, and Vavilov's immediate boss, Nikolai Gorbunov.

The newspapers hailed the conference as a triumph, but it was merely the beginning of another successful year for Vavilov. Now forty-two, he would become the youngest-ever full member of the USSR Academy of Sciences. Also elected was his former professor at the Petrovka, Dmitry Pryanishnikov, twenty-two years his senior. Vavilov's administrative responsibilities would be extended over a new all-encompassing Lenin Academy of Agriculture, based in Moscow, eventually giving him control of some 111 research institutes and 300 experimental stations across the Soviet Union. And yet he still found time for another expedition—this time to China and Japan. Behind the headlines, however, there were troubling signs of the bloody confrontation in biology that was to come.

The first, and strangest, of these signs was on display during Vavilov's Leningrad conference at the start of the year. Anyone attending also had an opportunity to view a Soviet propaganda film showing at a local Leningrad movie house. Entitled *Salamandra*, it was about a young biologist from an unnamed central European university who worked with salamanders and had successfully changed their color from brown to black by interaction with the environment. One day, this young biologist achieves the supposedly impossible. The newly acquired color is inherited in the salamander's offspring, proving the Lamarckian theory of the inheritance of acquired characteristics.

A priest learns of the experiment and is concerned that it might confuse his flock about the truth of the creation. He sets out to discredit the biologist. First, he persuades the biologist to announce his amazing discovery to a formal meeting of the university. Then, the night before, the priest arranges for a lab assistant to inject a salamander with black ink and substitute it for the experimental one. At the formal meeting of the university, the biologist makes a brilliant speech about his experiment, produces the salamander and puts it into a glass of water. All the color runs out of the specimen, the audience bursts into laughter, the biologist is humiliated and disgraced, and is dismissed from the university as an impostor.

The biologist becomes a beggar, walking the streets—until a former Russian student of his, a young woman, finds him destitute in an attic. She immediately takes the train to Moscow and obtains an interview with Anatoli Lunarcharsky, the commissar of education (a real person played by Lunarcharsky himself). Lunarcharsky orders his functionaries to save the biologist from bourgeois persecution. Meanwhile, the biologist is so miserable he decides to commit suicide, but the Russian student returns in time to stop him taking his own life. In the last scene the biologist and the Russian woman are in a train riding east and a large banner reads "To the Land of Liberty"—where, the viewer is supposed to deduce, Soviet science welcomes all researchers with contributions to the great advances of socialist science.

The story was based on the life, and suicide, of Paul Kammerer, an Austrian biologist who claimed that he had changed a key characteristic of a male toad by changing its mating habitat. Most toads, and frogs, mate in water. The male toad clasps the female around the waist and keeps her in place until she spawns her eggs, which he then fertilizes with his sperm. To get a grip on the female's slippery body, the male, in the mating season, develops rough black pads on the inside of its palm, known as nuptial pads.

Another type of toad, the so-called midwife toad, mates on land and does not possess these pads. Kammerer claimed that by forcing a male midwife toad to mate in water it had developed nuptial pads—and that these pads had been inherited by successive generations. An American biologist went to Vienna, examined the said toad, and found the dark color on his palm was caused by the injection of black ink.

In the scandal that followed, Western scientists, especially Vavilov's friend William Bateson, ridiculed Kammerer's experiment, and disgraced and in shame, Kammerer shot himself. But not, apparently, before he had arranged to move his laboratory to the Soviet Union, where he had been assured he could continue his experiments and might have become a Soviet hero. The film was Commissar Lunarcharsky's attempt to show that Western scientists were either steeped in clerical nonsense about creation or uncritical adherents of Mendelism, and they would do anything to discredit Lamarck's theory of the inheritance of acquired characteristics.

At the time the film was made, Lamarckism still found a sympathetic audience among traditional Russian botanists who considered, as Lunarcharsky apparently did, that classical genetics was insufficient to explain evolution. These botanists believed, among other things, that small mutations from prolonged exposure to environmental factors could be inherited. The vernalization experiments of Trofim Lysenko, where he talked about "training" plants in different environments, fit into the Lamarckian mold. In addition, as Soviet science developed, Lamarckism's emphasis on the interaction between the organism and the environment fit the general dialectical view of nature: that the organism's development should be regarded as an interaction of the whole and its parts, not governed by stable immutable genes, as they were being portrayed by the early geneticists.

The year 1929 would also mark the beginning of a sharp dispute in Soviet biology, and especially in genetics, between those who believed in Lamarckism and the geneticists who argued the reality of the gene as a hereditary substance. This basic difference would become the main issue of all following biological debates. Lamarckism had an intuitive appeal to practical breeders, the "gardeners" with little theoretical training in biology like Luther Burbank, Ivan Michurin, and Lysenko. In the context of Stalin's "Great Break with the Past" it did not help the geneticists that they relied, for the most part, on research done abroad—in Europe and America. Over the next few years, as the debate hardened, Vavilov would represent the geneticists and Lysenko the Lamarckists, or neo-Lamarckists, as they would be called.

In 1929 Vavilov was open to all points of view. While he did not believe in Lamarckism—there was no proof—he was not prepared, like some members of his staff, to shut out researchers like Lysenko whose experiments were based on the possible influence of environmental factors. Vavilov insisted that Lysenko should not only be invited to the Leningrad conference but also be allowed to give a paper. It was an honor and a big test for Lysenko. For the first time, he was to perform among his learned peers, the bourgeois specialists.

Three papers on methods of treating seeds to a period of cold—vernalization—were scheduled, one by Vavilov's expert, Nikolai Maksimov, another by the cotton specialist and close friend of Vavilov's, Gavril Zaitsev, and a third by Lysenko, co-authored with a collaborator. The main lecture was given by Maksimov. Zaitsev never made the conference. He died from peritonitis on the way.

In the event, Maksimov stole the show. The great problem of Soviet agriculture, he said, was the destruction of winter wheat, but research at Vavilov's Institute on plant stages of growth had reached the point where it would soon be possible to direct plant development at will: breeders would produce plants that bore fruit, or seeds, "at any time we wish." The economic importance of these achievements "needed no further elaboration." In contrast to Lysenko, Maksimov considered that the number of hours of daylight was more important than temperature, and overall he viewed Lysenko's experiments only as a "confirmation and further development" of earlier results obtained by the German researcher Gassner, and others. Lysenko's experiments were not "a scientific discovery in the precise sense of the word."[1]

Under the headline, "Winter Wheat Can Be Turned into Spring Cereals," the Leningrad *Pravda*'s report of the conference did not even mention Lysenko, only Maksimov. Lysenko might have been content with being asked to the prestigious conference, and the opportunity to present a paper to international experts, but he was disappointed, even angry, at the cool reception to his work. His colleague D. A. Dolgushin would recall, "The pillars of science used the tried and tested method of fighting: they ignored Lysenko's paper."

It was Lysenko's first encounter with the enemy. "Returning from the meeting of geneticists, Lysenko understood that he had addressed himself to the wrong quarters. His discovery was of no use to the dogmatic followers of Mendel and Morgan," Dolgushin would say.[2]

But Lysenko had learned from the *Pravda* article in 1927 that in the new socialist state other options were open to him. He had already prepared a second line of attack. He had told his peasant father, Denis, about his vernalization method and Denis had stored two sacks of seeds of winter wheat under the snow—to vernalize them—and planted them on half a hectare in the spring. By early summer of 1929, they were showing signs of producing a high yield—up to three times the normal, it was said, initially—and a delegation of agricultural officials had descended on Lysenko's home village of Karlovka and confirmed the expectations of a bumper harvest.

The Ukrainian Commissariat of Agriculture was so impressed

Nikolai Vavilov in his institute office in Leningrad, 1935. On the desk is a
map of his "centers of origin" of cultivated plants, the inspiration for his
global seed bank, his greatest scientific achievement.

2

At home in Moscow, December 1916. Nikolai Vavilov with his mother, Alexandra Mikhailovna, and his brother Sergei (in a tsarist officer's uniform) on leave from the front during World War I.

3

4

The Vavilov family home in Middle Presnya.

Nikolai Vavilov with his exiled father, Ivan Ilyich, in Berlin, 1927.

Vavilov wearing the formal uniform of a student of the Petrovskaya Agricultural Academy for his student ID card, 1909.

Researchers of the Bureau of Applied Botany, where Vavilov was a trainee in 1911. Front row (left to right): Vavilov; Andrei Maltsev (herb specialist); Robert Regel, Vavilov's Russian mentor; Konstantin Flyaksberg (wheat specialist).

Vavilov with William Bateson, his mentor during his postgraduate work in England. The photo was taken at the Pushkin experimental farm (the former estate of the Grand Duke Boris Vladimirovich) outside Leningrad, during Bateson's 1925 visit.

8

The somber wedding photo of Vavilov and his student friend Yekaterina Sakharova, Moscow, 1912.

Vavilov with his son Oleg, who was born in 1918.

9

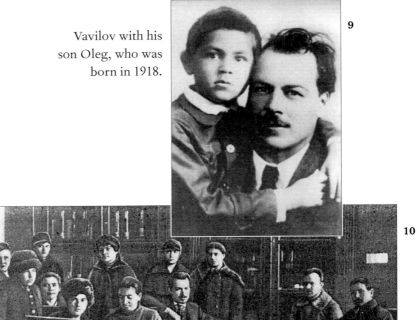

10

Vavilov with admiring students, mostly women, at the Faculty of Agronomy of Saratov University, 1920.

In 1921, Vavilov took over an empty tsarist palace with a grand staircase, gilded ceilings, and chandeliers, but no heat, in St. Isaac's Square, Petrograd. It would become the new home of the Department of Applied Botany and New Crops, later named the All-Union Institute of Plant Industry (VIR).

The research staff at the Department of Applied Botany, sorting plant specimens, 1923.

13

1924, Kabul, Afghanistan. Vavilov, in a rare moment without a hat, in the back seat of a Soviet embassy car as he sets off on his first major expedition.

14

1926. Vavilov, smartly dressed as usual, shakes hands with "Kazazmach," the leader of his plant-hunting caravan in Abyssinia.

15

Vavilov and his horse negotiating rocky terrain during one of his many expeditions in the Transcaucasus.

16 Vavilov at the Pushkin Experimental Farm outside Leningrad in 1925, with colleagues and his Soviet Communist Party patron, Nikolai Gorbunov, Lenin's former secretary. Left to right: Andrei Maltsev, arrested in June 1941 and released in the mid-1950s; Viktor Pisarev, head selectionist of the Pushkin farm, arrested in 1932 and exiled; unknown female secretary; Gorbunov, convicted as a spy and shot in 1938; Vavilov; the cytologist Grigori Levitsky, arrested in June 1941 and died in prison a year later.

17 Vavilov and Yelena Barulina after they revealed their secret love affair, 1926.

18

The house for researchers at Pushkin (modern view). Vavilov, Yelena, and their son, Yuri, had an apartment next to Georgy Karpechenko, plant geneticist and cytologist, and his wife, Galina.

A meeting at Vavilov's institute of experts on forage crops in 1935, showing one of the palace rooms with its hand-painted walls from which hangs
a large portrait of Lenin. Vavilov is in the middle of the seated row next to
his colleague, Yevgenia Sinskaya.

Vavilov went abroad, to North and South America, for the last time in 1932. He is seen here at the Sixth International Genetics Congress, Ithaca, New York, with the Nobel laureate Thomas Hunt Morgan (center) and the Russian geneticist Nikolai Timofeyev-Ressovsky, who was jailed in 1945 and sent to a labor camp.

20

21

On the same trip, Vavilov was in California on a mission to persuade two brilliant Russian geneticists to return home. Theodosius Dobzhansky (left, with his wife) declined the offer. Georgy Karpechenko (right), who went back, was arrested in 1941 and shot.

22

Vavilov brought back seeds from wherever he traveled. Here he is in 1932, collecting beans in a Mexican street market.

On an expedition in the mountains of Soviet Central Asia, 1936. Vavilov is in a hat and shirtsleeves, walking with other researchers ahead of their transport.

At the Near Aral Experimental Station of VIR, 1936. Vavilov is next to the pilot.

25

Trofim Lysenko, the "bare-foot scientist" praised by Stalin, examining an ear of wheat in Ukraine, 1934.

26

Isaak Prezent, Lysenko's political minder, visiting Ivan Michurin, the fruit breeder (right), at his Kozlov orchards, ca. 1934.

27

Stalin listens to a speech by Lysenko in the Kremlin honoring tractor operators, machinists, and workers after bringing in the best harvests, December 29, 1935.

28

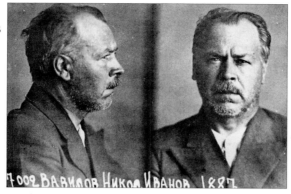

Prison photo of Vavilov, who was arrested by the NKVD on August 6, 1940, on charges of sabotage and spying for Britain. Interrogated and tortured, he was sentenced to be executed by firing squad.

29

After his conviction, using a scrap of prison paper, Vavilov appealed to Lavrenty Beria, the NKVD boss, to be allowed to work in a labor camp. Vavilov pledged to be "at the disposal of the Soviet power and my Motherland." He said he would "complete in half a year the 'manual on practical breeding varieties of cultivated plants resistant to major diseases,'" and "in six–eight months of strenuous work the 'manual on practical breeding of cereal grasses' applicable to conditions of different regions of the USSR."

30

After Vavilov's arrest, Yelena Barulina and Yuri, their son, stayed at Pushkin until the German invasion forced them to leave in the summer of 1941 (photo dated March 1941).

31

32

(Left) Saratov Prison, where Vavilov was incarcerated in October 1941 after being evacuated from Moscow. (Right) one of Vavilov's cells in Saratov Prison, where he died of malnutrition on January 26, 1943.

33

Sergei Vavilov, overburdened by his office as President of the Soviet Academy of Sciences, died of heart failure on January 25, 1951. He had made an unsuccessful appeal to Stalin to pardon his brother Nikolai. His picture hung in front of the Union House in Moscow, where the civil funeral was held.

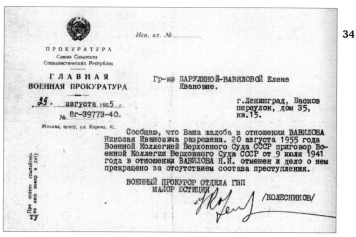

On August 20, 1955, the Military Prosecutor's Office rescinded the guilty verdict against Nikolai Ivanovich Vavilov. On August 29, this letter was sent to Yelena Barulina, the first official communication she had received in fifteen years, since Nikolai's arrest. The bloodless language reads, "I inform you that your complaint about the situation of Vavilov, Nikolai Ivanovich, has been resolved . . . the verdict against him is rescinded." Yelena Barulina died in 1957.

In November 1987, the Soviet Union marked the centenary of Nikolai Vavilov's birth with a huge gathering in the Central Concert Hall in Moscow and an outpouring of adulation in the press. A British diplomat wrote to the Foreign Office in London, "One more gap has been filled."

by Lysenko's demonstration at his father's farm that it ordered one thousand one-hectare tests for the spring of 1930. *Pravda* again gave Lysenko star billing. "The prospects opened up by the extraordinary discovery of agronomist Lysenko, corroborated by such outstanding experimental data, are so great that they are beyond immediate calculation," the article declared. A second article, ten weeks later, exclaimed, "The agronomist Lysenko has completely overthrown the definition of winter forms prevailing hitherto. ... Agronomist Lysenko's discovery will lead our agriculture onto a high road of vast possibilities and extraordinary achievements and greatly increase the tempo of our socialist construction."[3]

Such wild forecasts were nonsense in scientific terms, of course. A single experiment on half a hectare did not allow such a conclusion, and a host of other factors, some of them mentioned at the Leningrad conference, had to be taken into account—such as the problem of mold developing on the moist seeds, the difficulty of keeping the seeds at uniform temperature under the snow, transporting them to the farmers and sowing them in the fields without damaging them. But the editors of *Pravda*, intent on glorifying the revolution and the work of the "barefoot scientists," did not concern themselves with such fine technical details. Whatever Lysenko himself believed about his "discovery," he was increasingly willing to play his part as a socialist hero and set himself on a course of inevitable conflict with the older scientific establishment, and with his original promoter, a member of the middle-class professional elite, Nikolai Vavilov.[4]

Meanwhile, in an effort to increase political control, the Soviet government revamped the administration of agriculture. The presidium, or governing body, of Vavilov's Lenin Academy of Agriculture became a secretariat of administrators, not a committee of scholars, and it was overseen by a new Commissariat of Agriculture for the whole of the Soviet Union charged with carrying out Stalin's decision to collectivize the farms.

In public, Vavilov saw no contradiction in this arrangement; it was "two aspects of the same drive toward more efficient production."[5] He made all the right noises about the need for theoretical research to be "tightly linked to production." The researcher "must enter into the spirit of practical production." Farming was to be carried out in new-style collective work, like a factory production line. His new Lenin Academy was to be "the academy of the general staff of the agricultural revolution," the "general staff" being the new Commissariat of Agriculture.

However, Vavilov's new expanded duties inevitably clashed with his expeditions. He wanted to go to Japan and China, but Gorbunov, now secretary to the Council of People's Commissars, the head of the Soviet government, told him he would have to cancel his trip. Once again, Nikolai Ivanovich wrote a spirited defense of his expeditions, and once again he prevailed, as he had done over the expedition to Africa.

In early June, Vavilov departed for the Soviet Far East, Japan, and China, leaving Gorbunov to organize the new academy. In a whirlwind tour of western China, Japan, Formosa (now Taiwan), and Korea, he collected 3,500 kilograms of seed and herbarium samples and took several thousand photographs of the farming and geography. He did not return until December. He concluded boldly that the crop diversity in these countries was "wholly unique, that the cultivated plants were absolutely atypical, that the cultivation techniques were unmatched and that the ancient cultural center of East Asia was completely independent with its uncommon plants species and genera. . . .

"As yet hardly studied, the rich vegetation of China is known only from a few fragments in the reports of European and American travelers and undoubtedly preserves plenty of treasures . . . ahead of us is a vast expanse of work."[6]

But once again in his absence trouble was brewing. Stalin's "Great Break with the Past" was proceeding apace on the cultural front. Higher education was expanded to include students from appropriate proletarian backgrounds and at the end of June 1929, the party started work on "cleansing" the staff of the Academy of Sciences of its "class enemies."[7] A special commission set up to carry out this "cleansing" reflected the goal. The commission included only one scientist, the director of the seismology institute, a representative of the guards at the buildings of the academy, three workers from the main Leningrad factories, two other persons, and two representatives of the OGPU, Stalin's secret police and successor of the Cheka.

More than seven hundred of the academy's staff, mostly specialists in the humanitarian sciences, were dismissed based on their noble or bourgeois origin. In October, another special commission started to arrest academy members. Historians were charged with being members of a nonexistent "monarchist plot" under the name All People's Union of Struggle for the Restoration of Free Russia. One of their "crimes" was to have stored at the academy the Acts of Abdication of Nicholas II and his brother, Mikhail.[8]

Stalin called for a "determined offensive of socialism against the capitalist elements in town and country," especially the kulaks, the richest and the more successful farmers, a peasant bourgeoisie. The offensive was supposed to "eliminate the kulaks as a class." This would entail mass arrests and deportations of between five and six million people, or 4 percent of the peasant households. Any critical comment, or even silence, about Stalin's agrarian plan was grounds for dismissal, arrest, or even execution. Stalin predicted that by speeding up the creation of state farms, *sovkhoz*, mostly concentrations of big industries, like sugar, and the collective farms, *kolkhoz*, "there is no reason to doubt that in about three years' time our country will be one of the largest grain producers."[9]

Agricultural scientists would also be victims of the purge. The first was Alexei Doyarenko, a fifty-six-year-old professor at Vavilov's alma mater, the Petrovka. He was arrested for sabotage. He did not share Vavilov's optimism about the future of Soviet farming under Stalin's new policy. Soviet leaders faced enormous difficulties in feeding the sudden increase in urban workers caused by Stalin's fast-paced industrialization. The disastrous winter of 1927–28, during which five million hectares of winter wheat had perished, mainly in Ukraine, was followed by an even worse season in 1928–29 when seven million hectares of winter wheat were lost. The Soviet government was desperately searching for a quick solution. Doyarenko feared, correctly as it turned out, that Vavilov's enthusiasm was misplaced; his plant breeders would not be able to meet Stalin's expectations, and Stalin would blame them for the failure of collectivization. The viciousness of the assault on the older academics was seen in Doyarenko's arrest and eventual exile. A colleague who had sung at the recent funeral of Doyarenko's wife was also threatened with dismissal from the Petrovka for being too religious. His defense—that he was fond of music, not religion—proved worthless.[10]

Meanwhile, Lysenko continued to receive a favorable press. In the summer of 1930, he claimed that he had overcome a serious objection to his process of vernalization. The issue was whether the partially germinating seeds would go moldy before they were planted. He claimed that drying the seeds after the initial addition of moisture solved the problem. As long as the water content was below 30 percent the seeds would not go moldy. The claims turned out to be too optimistic.

By this time, Vavilov was becoming progressively less fascinated by Lysenko's speculations, but his response was still measured, at all costs trying to avoid a confrontation. There were no short-term solutions to the problem of the loss of winter wheat, he warned. "One must not shut one's eyes to the great difficulties, to the enormous amount of work that alone can lead to basic solutions of the most important tasks posed by life, posed by the disaster of the current year."[11]

At the same time, Vavilov was overwhelmed by the size of the new agricultural bureaucracy the government had created. He wrote to his colleague Georgy Karpechenko, who was studying with Morgan's fruit fly team in California. "I am completely squeezed here. On top of all the [Lenin] Academy work with its tens of institutes, I suddenly turned out to be my own 'supervisor' as I was made a voting member of the Collegium of the U.S.S.R. Commissariat for Land. We want to build Washington, nothing less. They told me, 'We will let you go in a year.' All in all, I have eighteen posts. My cranium will soon explode from these layers of rubbish on all sides."[12]

CHAPTER 15

State Security File 006854

At the beginning of 1930, the food shortages in the Soviet Union were so disastrous that Stalin ordered his secret police to find scapegoats. The OGPU, as the state security service was now called, invented a counterrevolutionary group that was supposed to have coordinated "wrecking" and "sabotage" activities in Soviet agriculture. Nikolai Ivanovich was listed as a leader of this group.

From the first days of the 1917 revolution, the secret police kept files on all prominent Soviet citizens, especially state officials like Nikolai Ivanovich who traveled abroad. Agents operated all over the world informing on the activities of Soviet citizens—and they spied on Vavilov as he moved through Europe, America, and Africa. In America in 1921, he was tailed by an agent code-named "Svezda." According to Vavilov's file, agent "Svezda" reported a conversation between Vavilov and the geneticist Thomas Hunt Morgan in which Vavilov is alleged to have said that the Russian people could not wait for capitalism to be restored. "Svezda" concluded that Nikolai Ivanovich was "an adventurist who put his own interests above the

Soviet state and built glory for himself at the expense of others." Other agents reported on Vavilov's contacts with Russian emigrants in Belgrade, Prague, and Berlin where Vavilov met up with his father and other Russians who had gone into voluntary exile. Agents also reported on Vavilov inside Afghanistan and noted that in Abyssinia he had met with a former general of the tsarist army.

On March 11, 1930, the OGPU opened Vavilov's Operational File No. 006854.[1] On that day, the file records that an anonymous inmate of an OGPU jail in Ukraine informed agents that there was an active counterrevolutionary cell inside Vavilov's Leningrad Institute. The inmate, who himself had been arrested on charges of agricultural sabotage, also pointed to a counterrevolutionary "experimental group" in Moscow. Agents started compiling a dossier accusing Nikolai Ivanovich of attempting to sabotage Soviet agriculture. The dossier would eventually contain 136 files, and would form the basis of his arrest and brutal interrogation in 1940.

By the end of 1930, the OGPU had "established" that Nikolai Ivanovich was the head of this anti-Soviet group. In a particular incident, the agents reported, there was evidence that Academician Vavilov had supported Tashkent workers "resisting the expansion of cotton growing in new areas."

The presence of Vavilov's anti-Soviet ring would be "confirmed" by "the recruited special agent Kalchuk [a code name]," who "used to be close to the group of Vavilov, but was expelled from it for treason." Kalchuk presented "abundant materials about the wreckage plan of the ring." For the next ten years, the OGPU would receive reports of Vavilov's counterrevolutionary activities from other detainees and paid secret police informers inside the Institute.

This was how Stalin's terror system worked. Under Stalin's orders, the OGPU gathered what it called "evidence" on his political rivals, those he decided were "wreckers" or "saboteurs" or "enemies of the people," and those he could blame for the failure of his policies. This "evidence" was often obtained under torture, or in exchange for becoming an informer. The "evidence" would be made public in the press at an appropriate political moment. Arrests and a show trial would follow.[2]

In the case of agriculture, the OGPU started to fabricate this evidence in the spring of 1930. The targets were the agricultural scientists and economists, the plant breeders, and the academics in agricultural institutes.

In September 1930, in the midst of another poor harvest, the newspapers announced that the OGPU had uncovered a counter-revolutionary organization named the Peasant Labor Party (TKP).[3] Membership of the TKP was claimed to be 100,000 to 200,000 and the group was said to have an extensive network in various government organizations.

The TKP was entirely fictitious, but in the next few months, thousands of its "members" were arrested. Nikolai Kondratiev, a renowned agricultural economist, was sentenced to eight years in prison. Alexander Chayanov, another economist, was sent into exile in Kazakhstan. Many officials in food production; including former owners of large estates, confessed to sabotage intended to create famine, and to being agents of imperialism. According to the newspaper, among other crimes, they were responsible for the decline in meat production. After a short trial, they were all found guilty and sentenced to death without right of appeal. Forty-six other agricultural experts and scientists were executed almost immediately after the trial.

Some of those arrested traded their freedom to become an OGPU informer against Vavilov. One of them was Ivan Yakushkin, also a professor at the Petrovka. Yakushkin was of noble birth. After the 1917 revolution, he could not tolerate the Bolsheviks and tried to leave Russia through the Crimea with the retreating White forces.[4] But at the last minute he was thrown off a boat bound for Turkey and was forced ashore. He eventually got a job as a research worker on a sugar beet experimental farm in Voronezh, three hundred miles south of Moscow. He was arrested in 1930 as a member of the Peasant Labor Party but released a year later, and recruited as an informer. When he returned to his job at the Petrovka, he began sending reports about his colleagues to the OGPU—"particularly about Vavilov," according to Vavilov's dossier.[5]

Yakushkin had known Vavilov since student days. His sister Olga had studied at the Petrovka with Vavilov. In September 1931, Yakushkin delivered his first ten-page report to the OGPU. He accused Vavilov's Institute of being at the center of anti-Soviet activity and charged Vavilov personally with being the organizer of "wrecking activities" in plant selection and seed production. The fact that everyone familiar with Vavilov knew how dedicated he was to his science did not deter Yakushkin. Like others sent to spy in this terrible period, Yakushkin's continued freedom depended on telling Stalin's state security what they wanted to hear.

Another secret collaborator was the plant breeder Alexander Kol. A former employee at the Leningrad Institute, he had quarreled with Vavilov over research policy. Kol, who was ten years older than Vavilov, considered himself a superior academic and was very ambitious. He had argued that the scope of the world collection was too broad and should be confined only to those plants that could be introduced directly into farm production. This issue was at the core of Vavilov's global vision. Other countries, notably the United States, where Kol had worked briefly, had concentrated on a few cultivated plants, especially wheat and maize. Vavilov wanted to collect all kinds of cultivated plants and make his world collection the most comprehensive—even though, as he admitted, it would take time. "It is as clear as day that there is enough work in the sphere of [plant] genetics proper for decades for dozens of people," Vavilov wrote in early 1930.[6] "There is enough [work] for a whole generation, for hundreds of researchers, but the engine has to be started."

Kol began to quarrel openly with Vavilov over this issue, but Vavilov stood firm, arguing that it would be extremely shortsighted to cut back on theoretical research. Vavilov had his supporters at the Institute, some of whom regarded Kol as a troublemaker and a careless worker. When sending seeds back from his expeditions, Kol would often lose them or their labels.[7] In trying to avoid a confrontation with a colleague, Vavilov began to move Kol's tasks to other departments and Kol complained he was being treated badly.

The dispute seemed to go away, but on January 29, 1931, Kol publicly attacked Vavilov in an article in the influential newspaper *Economic Life*, published by the Council on Labor and Defense, a government super-cabinet. The article was a startling attack accusing Vavilov of being a counterrevolutionary. "Under the cover of Lenin's name a thoroughly reactionary institution, having no relation to Lenin's thoughts or intents, but rather alien in class and inimical to them, has been established and is gaining a monopoly in our agricultural science. It is the Plant Breeding Institute of the Lenin Academy of Agricultural Sciences." Kol specifically attacked Vavilov's expeditions. "Work . . . on the centers of origin of plants has now replaced V. I. Lenin's revolutionary task of invigorating the Soviet farmland with new plants."[8]

Kol denounced Vavilov's plan to build the world's largest collection of cultivated crop seeds as "reactionary botanical studies" that had resulted from Vavilov's obsession with his Law of Homologous Series and his separation of "theory and practice."[9]

The newspaper endorsed Kol's article in an editorial, thus expecting, in the normal way, an apology or at least self-criticism from Vavilov. In his reply Nikolai Ivanovich conceded that the new Lenin Academy—shortly to have 1,300 establishments, ranging from institutes to small experimental stations, and 26,000 employees—was engaged in a grand vision and could be intelligently criticized for the enormous sweep [*razmakh*] of its operations, but he insisted that Kol was only attacking the academy and his Leningrad Institute because of his personal issues.[10] What Nikolai Ivanovich did not know was that Kol was about to become an important OGPU informer, if he was not one already. Kol would be arrested in the Peasant Labor Party crackdown and quickly released. His ticket to survival was to become a collaborator and assist in the destruction of his former boss.

Kol's article was the first public attack on what opponents began to call Vavilov's "Institute of Noblesse Botanists." Several critical articles followed calling the Institute's work too theoretical, and complaining that it was staffed by the old guard of specialists.

Vavilov's OGPU dossier also reveals that a third former employee of the Institute became a secret collaborator. He, or she, is identified only by the code name "Vinogradov." This person had been arrested as a counterrevolutionary and later confessed that a group of specialists headed by Vavilov was "carrying out a fight against the socialist reconstruction of agriculture." Later in 1931, a fourth informant reported, "From the very first years of the existence of Soviet power, the selection [of plants] business found itself in the hands of a small group of the old mold, closely connected with classes hostile to the proletariat revolution. . . . Vavilov was the head of this section [which] was used for wreckage purposes and has seriously slowed down the reconstruction and development of the Soviet plant breeding."[11]

The OGPU dossier that was opened in the spring of 1930 was building silently. In the late summer, Vavilov sailed off to another conference in America, where agent "Svezda" resumed his surveillance and created more of his warped reports for his masters in Moscow.

CHAPTER 16

The Passionate Patriot

*As Stalin's secret police opened their file, Nikolai Ivanovich was
actually engaged in acts of great loyalty to the state, hunting
for plants of strategic interest such as rubber and quinine. And
he was trying to persuade the best Russian geneticists, working
temporarily in America and Europe, to return immediately to
join the great socialist experiment.*

In the fall of 1930, Nikolai Ivanovich attended two international
conferences of agronomists, one in Ithaca, New York, and the other
in Washington, D.C. Then he went plant hunting. He moved quickly
through ten of the leading farm states, including California, and
even managed a side trip to Mexico, in part to sample the country's
unique diversity of maize. Mexico was a true "center of origin," he
declared after sampling its varieties. But Mexico had a second, less
academic attraction: guayule (pronounced why-you-lee), a small
shrub containing gooey latex, similar to the sap of the rubber tree.

In the 1920s when leaf blight hit the Brazilian rubber industry,
guayule was briefly the subject of intense agricultural research. The
rapidly expanding industrial nations were searching for new and

cheap sources of rubber that the British, who dominated the rubber market, could not control. At the end of the nineteenth century, a British plant hunter had smuggled seeds of the rubber tree out of Brazil and the British now ran so many rubber plantations in their Far East colonies that they had eclipsed Brazil's production. Guayule seemed a possible source of rubber or a rubberlike substance, at least, for the Russians, and Vavilov, knowing the Mexicans had little interest in the plant, began collecting specimens in his usual frenetic way. He could not have known that he was about to cause an international incident.

The Americans were also looking at the possibilities of using guayule, and had even produced viable quantities of latex from guayule.

Communists were stealing Mexico's treasures, the Americans howled, rallying the Mexican press into protesting the "plundering of national treasures by the Bolsheviks."[1] Even so, Vavilov managed to bring guayule seeds home, but the first sowing in Azerbaijan was a disaster. Ninety percent of the plants were killed by frost.

Apart from seeds, Nikolai Ivanovich had a personal mission. He wanted to persuade two younger, talented Russian geneticists working in America to return immediately to Leningrad and join his plant breeding enterprise. The two researchers were working in Thomas Hunt Morgan's famous fruit fly laboratory, which had moved to Pasadena. One was Theodosius Dobzhansky, a brilliant zoologist who had worked closely with Vavilov in Leningrad. The other was Georgy Karpechenko, who had been awarded a Rockefeller Prize for producing a fertile cross between a radish and cabbage. This creation was thought to be impossible because the radish and the cabbage are from different genera.[2]

Dobzhansky and Karpechenko were thrilled to be in Morgan's lab at the forefront of international genetic research. They were enjoying the comforts of California, far away from the cold and the class wars raging in Leningrad. They could devote their time to pure science instead of defending themselves as bourgeois specialists at odds with the new Soviet culture. That Vavilov should even think of trying to lure them back with all the troubles brewing around him and his Institute seems, in retrospect, an act of dogged naïveté. The wonder is that Vavilov did not decide to join them.

His efforts to recruit his colleagues provide an important insight into Nikolai Ivanovich's unshakable faith, blind faith as it turned out, in the Soviet experiment taking over his beloved Russia. Vavilov was a Russian patriot, much like his father, Ivan Ilyich, before the revolution. Nikolai Ivanovich was convinced that Russians could be leaders in world science given the chance, and he believed the Bolsheviks would ultimately provide this opportunity. He simply ignored the appalling hardships of Stalin's regime—the shortage of funds, housing, food, medical care, the enforced ideology, and the terror of the secret police. He managed to set aside such trials in the higher interests of building what he believed would be the best, and purist, science in the world.

Vavilov wrote his two friends encouraging letters about the new opportunities in the Soviet state.

"Work is full steam ahead," he told Karpechenko. The monograph on wild oats has appeared, [Alexander] Maltsev has assured himself of immortality by it. It is a book of which we can be proud. We have published a volume on fruit growing, dealing with the wild fruits of the Caucasus, central Asia and the Far East. You can recommend it to anyone . . . it is full of original material . . . [Dmitry] Borodin's book is out. It has a very important article by [Nikolai] Kuleshov on maize and its worldwide distribution. Advise your comrades in America to have a look at it—it will do them no harm."[3]

Vavilov constantly prodded his Russian colleagues to share their knowledge of America. "Tell us about the wonderful things being done in America and take the best," he wrote, "We need everything that is good. We want to catch up at all costs."[4]

Vavilov either did not see or refused to recognize what so many others knew—that the regime was increasingly based less on reason and community than on terror; that the compact Lenin had made with the bourgeois scientists was under growing threat from the new cadres of the *vydvizhentsy*, the barely educated pushed-ups like Lysenko. Vavilov expected his colleagues to think like him, to see the same distant goal for their country. He coaxed, he badgered; he even mocked them to make his case. At one point, when Karpechenko failed to send the reports required, Nikolai Ivanovich chided his colleague, "Karpechenko who sits free and in the God-blessed climate of California while Vavilov [is] so burdened by editing research manuscripts and organizing the task of the Academy that he grabs only hours, rarely days for his [scientific] work."[5]

Karpechenko would finally succumb to Vavilov's taunts. He returned to Russia the following year to work at Vavilov's side, an act of loyalty to his science and his colleague that would have a tragic end. After he organized the plant genetics department at the University of Leningrad, Karpechenko then ran the genetics lab at Vavilov's Institute. He would become Vavilov's faithful ally in the coming battle with Lysenko, but eventually, like Vavilov, he would become another casualty of this growing scientific and human disaster. In the final purge of the geneticists that would destroy Vavilov, Karpechenko would also be arrested and eventually executed for his pursuit of un-politicized science.

Karpechenko's colleague, Dobzhansky, would remain in America, despite Vavilov's entreaties. Many years later, Dobzhansky would describe how Vavilov, the "passionate Russian patriot," as he called him, first approached him to return.[6]

It was in October 1930, during a walk in the Sequoia National Forest where the two botanists had gone to be alone and avoid their personal conversation being overheard by unfriendly ears.

"Outside Russia Vavilov was considered a Communist, which he was not," Dobzhansky began. "But he accepted the revolution wholeheartedly because he thought it would open greater possibilities for Russia's people. . . . Vavilov said with great enthusiasm and conviction that in his opinion the possibilities for satisfying human needs in the U.S.S.R. were so great and so inspiring that for their sake alone one could excuse the cruelty of the regime. He asserted that nowhere else in the world was the work of the scientist valued as highly as in the U.S.S.R."

Vavilov followed up their conversation in the woods with an intense barrage of letters—first appealing to Dobzhansky's patriotism, then trying to dispel his concerns about living conditions, salary, and the rise to power in Moscow of those who would skew scientific policy and research. But Dobzhansky did not feel the same intense passion for Russia and could not ignore the cruelty of the Bolsheviks.[7]

Nikolai Ivanovich began to show irritation at Dobzhansky's hesitation.

"Feodosy Georgievich, help us lift the country," one letter began. "This is a mission for all humanity. And start acting as a serious Soviet patriot. Truly, the horizon here is wider and the future far more reliable than the comfort yet insecurity [of America]."

Vavilov's faith was seemingly unshakable. "Certainly the future belongs to us. It should be clear as two times two to our world-wandering kin [such as Dobzhansky and Karpechenko]. . . . The main point is that there is more good than bad in the world and in people. It's better to see the good, not the bad."

When Dobzhansky still wavered, Vavilov berated him.

"I personally find it strange to hear your queries about such practical matters. We are not used to paying so much attention to them: the same kind of people live here [in Russia, as in America], with two feet and two hands, not better and not worse than the Americans."

Vavilov conceded that scientists were paid "little," but they were "in a fairly decent situation . . . perhaps the housing question is going to be the most difficult so start approaching people about it beforehand. As a member of the government committee I have some idea of this business. Anyway, in the first weeks you'll be getting the same food ration as workers. We are building our life . . . our country is under construction and there is no doubt that in 3–5 years none of us will recognize it . . . that's why even the top theoreticians are expected to offer to help."

In any case, Vavilov added, Dobzhansky was a brilliant scientist and he could make extra money writing books. "With your talent for writing you won't find any difficulties. There is a need for books now, they are printed eagerly as long as they are interesting. Start writing now so that you have a fund by the time you arrive. We will help you publish."

When Dobzhansky wondered how he would cope with Marxist ideology, with the new pushed-ups and their insistence that dialectics be applied to science, Vavilov dismissed his concerns. "Dialectic methodology is nothing but a plus, it allows one to stay in touch with life's demands. . . . Of course, you must horse-shoe yourself into dialectics. It's an easy business and will bring you nothing but usefulness. I am going to send you my dialectic masterpiece in a few days, 'Linnaean Species as a System.' Perhaps dialecticians will criticize me for it, but at least for me the dialectic approach was useful."

Vavilov even suggested to Dobzhansky that he should look up "a number of useful articles" in the Communist Party's theoretical journals *Under the Banner of Marxism* and *Natural Sciences and Marxism*. "You need to know it all then you will be armed from head to toe."

But Dobzhanksy was still not persuaded. He asked Vavilov to arrange for an extension to his allotted period abroad, but Vavilov told him bluntly that there was no chance because there was no budget for it.

"Come back and that's it," he wrote. "No need to burden people with your individuality, attracting too much attention. . . . You sat overseas and worked, published your papers in the Soviet Union, did a lot of good research, and so be it. No harm done . . . [but] you've got to cross the Rubicon."

Dobzhansky wisely did not return. In his last letter to Vavilov, he wrote, "Forgive me, Nikolai Ivanovich, if I hurt you. But I am writing to you honestly. With all my respect to you personally, with all my sincere desire to work at the Academy of Sciences, not here (I know that many doubt the sincerity of my desire, but this is up to them, I am saying what I think). I see that it is not going to work. It wasn't easy for me to write this letter, but it would have been worse not writing it. If I can ever be useful here, I will do my best. No matter what, I shall never forget the country and what I owe her."[8]

Nikolai Ivanovich was obviously disappointed; his irritation with Dobzhansky was a side of his character rarely seen by the public or even his staff. He would often go out of his way to avoid cross words, especially with close colleagues. And he often withheld criticism, a tactic that would work to his disadvantage in the growing scientific battles with Lysenko.

This was a lonely time for Vavilov. He was beginning to understand how burdensome the fight would become if some of his most talented colleagues were not there to support him. But he had chosen this route, a solitary and difficult path that balanced a faith in the communist experiment without his being a member of the party. At this point, his clashes with the party had been few—and manageable—but the threats to his independence were increasing.

An incident on this second trip to America illustrates the party's rising power. He was in northern Arizona visiting the Navajo and Hopi reservations to examine varieties of native corn when he received an official telegram. The wire from Moscow ordered him to appear at a state dinner in Washington the following week when the U.S. secretary of state and other leading American officials would be present. Ten days later, there would be another state dinner—in London at which the prime minister would be present. It was "essential" he attend both events.[9]

At the time he received the cable he was staying with Homer Shantz, president of the University of Arizona, and Shantz would later recall, "This was what he most dreaded. He was eager to complete his trip. . . . We had been together for some days and had many talks about the future of agriculture . . . and of the necessity of providing food for the then starving people of Russia and elsewhere. His whole desire was to improve agriculture and thereby raise the standard of nutrition for his people and the rest of the world. His trip into Mexico and Central America was of the greatest importance."

Vavilov hesitated. "At last he said, in essence, 'If I were a Communist I would have to obey and in that case I would not be able to use my own judgment. I am employed by the Communists to work for the welfare of the people of the USSR, but I am still free to judge what is best.' In the end, he refused to go to Washington or London, saying that his plant hunting was more important for the welfare of Soviet citizens than any state dinner could possibly be."[10] It was the right move for the pursuit of science, but it was not the right time for even a talented scientist in Stalin's Russia to resist the party's call.

CHAPTER 17

A Modest Compromise

In the early 1930s, young ideologues began graduating from the communist academies, their heads filled with Marxist theory, but not much science. At Vavilov's Institute they began attacking the research program for being too theoretical. Nikolai Ivanovich agreed that science must be made relevant to the country's needs, but he was not prepared to abandon his seed bank, the core of his dream to feed the world.

Vavilov returned from America at the beginning of 1931 to find the *vydvizhentsy*, the pushed-up researchers, trying to take control at his Institute. They wanted to close down parts of the Leningrad operation they considered too theoretical and hand over the work of collecting new strains to the branch institutes elsewhere around the country. Vavilov argued passionately against what he saw as nothing less than the destruction of Russia's agricultural future.

The suggestion "by some learned comrades" to hand over the work to branch institutes was "absurd" because "only a central institution, manned by specialists" could carry out this important task," he argued in a letter to the agriculture commissar, Yakov Yakovlev.[1]

"One may disagree about the principles and subject them to discussion, but unfortunately matters have gone further than this. In fact, there happens daily some kind of open or clandestine action to change parts of our working program. Only the return of the director from abroad has somewhat slowed the tempo of these events."[2]

The performance of the Institute was threatened, he continued, by "the easy-mindedness of a number of Party comrades, poorly trained but infected with a combination of criticism and desire for reform." The events had "forced" Vavilov to wonder whether he should stay on as head of the Institute.

In addition, the Institute was seriously underfunded and operating in a "totally abnormal condition." There was not even heat—only two degrees [Centigrade] indoors in the winter. The Institute had outstanding scientists, but they were paid below normal rates, even lower than employees in the branch institutes."

Much as he would have liked to avoid the matter of the young ideologues altogether, Vavilov was forced to deal with them. In his view, they were ruining Russian agriculture. But their Marxist rhetoric was a symptom of a wider problem. Vavilov realized that it was no longer enough merely to talk in scientific terms about his obsessions—the origin of cultivated plants, his expeditions, his seed bank and its link with the future of world food supplies. A reference to Marx, or Marxism, in lectures and speeches was "a demonstration of loyalty in these times and a reference to the practicality of scientific research was a pledge of usefulness."[3] Using Marxist rhetoric became "a negotiating language" employed by scientists to legitimate and justify their research agenda—and institutional ambitions—in the eyes of state officials.[4] Vavilov had never taken courses in Marxist theory, but he was prepared to "horse-shoe" himself into the dialectical method, as he had described the process in a letter to Dobzhanksy.

That summer, there was an international conference in London on the history of science and technology entitled "Science at the

Crossroads." For Moscow, the conference was a chance to gloat at the crisis facing capitalism during the years of the Great Depression. The Politburo sent a large delegation, including Vavilov and Nikolai Bukharin, the party's chief theorist.[5] Bukharin had fallen out with Stalin over the headlong rush into industrialization and collectivization, favoring a slower pace. Bukharin and his followers were expelled from the Politburo as "right-wingers," but Bukharin believed, as much as Stalin, in the need to unite theory and practice in agriculture—it was a key difference between the capitalist and communist approaches to science.[6]

The Soviet delegation flew into London on a special plane—with Vavilov managing to sneak his son Oleg on board at the last minute. Oleg, now almost twelve years old, was an extremely bright child who spoke some English. While Nikolai Ivanovich was at the conference, Oleg toured London on his own. Each Soviet delegate was given $200 for expenses—"Too little for the two of us," Nikolai Ivanovich wrote to Lenochka, "but we'll try to make out."[7]

At the conference Bukharin gave a Marxist explanation of the "most profound crisis" in capitalism, "a crisis in individual branches of science, in epistemology, in world outlook, in world feeling." He spoke about the Marxist "primacy of practice"—that scientific truth was dependent on "practical" social aims, and that a scientific theory was only "correct" if it led to practical success. Science should represent the interests of the working class, he declared; the aim of proletarian science was to change the world, not to understand it.

The prime example of true revolutionary unity between theory and practice was Soviet agriculture, Bukharin continued. "One can feel with one's hands how the development of socialist agriculture pushes forward the development of genetics, biology generally, and so on," Bukharin continued.[8]

In his speech, Vavilov agreed with Bukharin that it was necessary "to break with the old academism," that the days of "scholastic monasteries, the laboratories of the alchemists and the quiet offices of individual university scholars" were over. He even agreed with the young ideologues that the *zamknutyi*, the academics who were "locked up" or "out of touch," should be forced to change their ways and that the ultimate aim was to "master the historical process."

But Vavilov concentrated on his own contribution to socialist agriculture—his collection of seeds from around the world. His way of uniting theory and practice was to use his world collection to "direct the evolution of cultivated plants and domestic animals according to our will."[9]

In a carefully worded paragraph that seemed to be aimed more at the *vydvizhentsy* in Leningrad than the international group in London, he said, "By knowledge of the past, by studying the elements from which agriculture has developed, by collecting cultivated plants and domestic animals in the ancient centers of agriculture, we seek to master the historical process. We wish to know how to modify cultivated plants and domestic animals according to the requirements of the day." He was only "slightly interested in the wheat and barley found in the graves of the Pharoahs of the earliest dynasties. To us constructive questions—problems which interested the engineer—are more urgent."[10]

His message was clear. He would make the right political noises, but he was not willing to abandon theoretical genetics. If only the peasant farmers would study his pamphlets, if only they would learn to use his Mendelian breeding methods properly, then they could show the very results that Stalin and the Politburo were demanding. The tools for reform and improvement were available; what was required was simply for the scientists in the research stations to teach the peasants how to use them.

If there was one central problem in Vavilov's grand plan, this was it. The resources to create the kind of massive extension service he had in mind—even bigger than America's—were simply not available in the Soviet Union. The funds were inadequate and the staff unqualified.

• • •

Meanwhile, Stalin continued to enforce collectivization, and, to ensure supplies of food for the factory workers, brought grain reserves from the south into central collection points for distribution to the hungry cities. This left rural areas without enough grain from the year's harvest. As summer turned into late autumn of 1931, Vavilov feared disaster. Another famine was imminent, this one man-made.

As the food supply for the rural areas began to disappear, the government blamed the plant breeders, calling for faster results of new varieties. At first, Vavilov responded with optimistic forecasts. "There is no doubt whatsoever that in the shortest possible time this new system will eliminate the discrepancy between research and the demands of life that our agriculture faces," he assured the authorities.[11]

But Moscow needed more than words. If the citizens were not fed, the revolution itself was threatened. At the beginning of August 1931, the government issued an astonishing, and totally impractical, demand. The normal ten to twelve years required to develop new crop varieties were to be reduced to four years. For wheat, the demands were even tougher. The Soviet leaders ordered that low-yield seed now being used by the farms must be replaced by new, certified high-yield varieties, *within two years*. Wheat was to be improved so that it could rapidly replace rye in the north and east. The decree also affected potato production. Newer varieties were to overcome diseases common in regions with hot dry summers and these varieties were to be produced in four to five years, instead of the normal ten to twelve.

This feat, the government declared, would be possible because of the Soviet system of collectivization—"because of the state farms and the collective farms and the planned system of directing agriculture." Of course, the government added, this would be impossible "in any of the capitalist countries, even the most agriculturally advanced"—because they had the wrong system.[12]

Importantly, the government was not asking Vavilov to drop his genetics research; the August 1931 decree was to be accomplished using "new foreign technology and the newest improved methods of breeding (based on genetics)." But the overall effort was no longer to be confined to scientific institutes. "The broad mass of collective farmers and workers on state farms must be drawn into this work."[13]

The order was impossible to carry out, as Vavilov and his colleagues knew, even using hothouses, as the government had also suggested. Growing plants in hothouses would indeed speed up production of new varieties but those varieties would still need to be tested for the widely different environments. Vavilov asked Commissar Yakovlev for more time, saying the best they could do was to produce the new varieties in ten, not twelve, years, but Yakovlev dismissed his pleading. "We do not have ten years to wait," he said.[14]

Lysenko saw opportunity in the government's decree. Without providing any experimental data, he would announce by the summer of 1934 that he had increased the yield on Azerbaijan wheat grown in Odessa by 40 percent using a variation of his method of vernalization—changing the temperature while the plants were growing. Commissar Yakovlev seized on the announcement as more evidence that Lysenko, not Vavilov, was the scientist who would answer Stalin's demand for the criterion of practice.

If Vavilov felt the lengthening shadow of Lysenko in any way menacing, or that Lysenko's speculative achievements threatened his own position, he did not show it. Quite the contrary, he continued to praise Lysenko's work as complementary to his own. Vavilov was producing results based on the theory of genetics; Lysenko was experimenting with the influence of the environment—temperature, light, and moisture. There was no scientific reason why the two methods should not exist side by side. He argued that vernalization would be important in breeding new varieties from his world collection—but he saw "no valid basis" for using it so soon in broad production experiments.[15] But Commissar Yakovlev had already made up his mind. The government ordered mass trials of vernalized crops in several regions of the country.

CHAPTER 18

The Red Professor

Vavilov and Lysenko coexisted in an uneasy truce until Lysenko started working with a party ideologue named Isaak Prezent. Under Prezent's guidance, Lysenko's speculative claims increased.

A former political commissar in the Red Army, Isaak Prezent had a sharp mind, a vicious tongue, and a deceivingly pleasant face. He was an extreme example of the power of the *vydvizhentsy*. He began his academic career as a lecturer at Leningrad University, teaching the new, politically correct ways of interpreting theories in biology. In reality, he knew very little about biology, having studied as a lawyer and social scientist. He even prided himself, as Lysenko would soon do, on not having read the latest research on genetics. He delighted in being a class vigilante and especially enjoyed bullying students of the bourgeoisie who knew genetics, but no Marxist theory. His lectures had such titles as "Class Struggle on the Natural Science Front," in which he attacked traditional liberal ideals of science. He once tried to get a young female student expelled for not being able to speak in class about the Marxist-Leninist "theory of cognition."[1]

While still in Leningrad, he started writing political pamphlets criticizing his fellow teachers. In one of these pamphlets, he attacked a Leningrad teacher who had written a harmless poem about the joys of the May Day holiday. Prezent castigated the teacher for not pointing out that this year was the thirteenth year of the revolution and that May Day was "a holiday of struggle, not of flowers."

From such petty encounters, Prezent launched personal campaigns against older bourgeois professors. He ridiculed the idea of a "pure science" independent of political loyalties and immediate practical usefulness. In one campaign his target was Boris Raikov, an outstanding Darwinian scientist twenty years his senior. Prezent denounced Raikov as an "agent of the world bourgeoisie, who would arouse nothing but loathing, disgust and hatred in every honest comrade."[2] Raikov was arrested in 1931, tried, exiled to the north, and was only allowed to return to Leningrad at the end of World War II in 1945. The term "Raikovism" was adopted as a method of condemning politically unacceptable scientific trends, and Prezent had established himself as a political force. Those who encountered Prezent at this time recalled the young man, in his early twenties, prancing about in front of his audience, waving his arms when making a political point.[3]

Within a few years Prezent would become Lysenko's political minder, promoter, publicist, and ideological spin doctor. Without him, Lysenko might have remained a lowly plant breeder with a good eye for a fine variety of wheat and several, speculative, anti-science ideas about how to improve crop yields. But with Prezent by his side Lysenko would become increasingly daring, and his scientific claims increasingly fraudulent as he worked to give the government, and even Stalin himself, exactly what they demanded. Lysenko offered quick solutions to the entrenched problems of Soviet agriculture. Prezent would broaden Lysenko's narrow political vision, feed his ambition, trigger his native cunning, and show him how a peasant farmer's son could become a political force in Moscow.

Lysenko said he first met Prezent in 1933,[4] but another version says they met earlier at the 1929 genetics congress when Lysenko put forward his idea of vernalization and was summarily rebuffed as a know-nothing by Vavilov's staff. One version of this meeting has Prezent approaching Lysenko after his speech and suggesting that it would be a more profitable course to link his vernalization method with a known authority such as Darwin. Lysenko asked Prezent who Darwin was, and where he could meet him.[5] (The story may have been concocted by Lysenko's enemies. But Lysenko himself would later admit he had never read Darwin and that he always relied on Prezent for an interpretation of Darwin's works.)

At any rate, two years later, in 1931, Prezent spotted a promising opportunity for himself as Lysenko's partner. By this time the Marxist philosophers and their tribunes were enthusiastically peddling their fusion of theory and practice doctrine, and their work had prepared the ground for the "drift" away from Vavilov's long-term plant breeding and toward any theory that answered their short-term needs and thus Stalin's demands.[6] The agriculture commissar, Yakovlev, had already publicly endorsed field trials of Lysenko's vernalization, praising its revolutionary significance for science as well as agriculture.[7] Prezent would soon move to Odessa, where Lysenko was setting up his operation. Their "creative partnership," as it became known, began. It was supported by its own media, *Iarovizatsia*, the *Journal of Vernalization*, edited by comrade Prezent. Lysenko would write that "Prezent got so closely intertwined with my work that . . . not a single new issue of the theory that we've been developing [was made] without detailed discussion and participation of comrade Prezent."[8]

Commissar Yakovlev was now so enamored of Lysenko's work that he instructed Vavilov give "every assistance" to Lysenko, and to "personally look after him."[9] Vavilov went to Odessa to see Lysenko's work for himself. The visit did not change his mind. He saw the advantages of using vernalization for plant breeding, but not on a mass scale for farming. Vavilov wrote back to his colleagues at the Institute that Lysenko's work was remarkable. It meant that a great deal would have to be put on a different footing, and the worldwide collection would have to be subjected to vernalization. But, he added, there were still a lot of mistakes.

Vavilov was disappointed in Lysenko's negative attitude to genetics but, generous as always, he thought this might be overcome if Lysenko got out of Odessa and came to Leningrad—even better if he went to an international conference. Vavilov invited Lysenko to attend the Sixth International Congress of Genetics and Selection, the five-year star event of genetics that was to be held in Ithaca, New York, in August 1932. Vavilov was the vice president of the congress and was hoping to bring a large delegation of Soviet scientists. His offer to widen Lysenko's experience seems to have been perfectly genuine although Lysenko would certainly have found himself out of his depth on genetics theory. It was Vavilov's way of patching over what he called the "disharmony" between his Leningrad researchers and Lysenko's Odessa group. He was even thinking of asking Lysenko to come to the Pushkin station several times a year. "It will be easier to do it under their guidance," he advised a colleague.[10]

In the speech Vavilov was preparing for the conference he would again refer to Lysenko's "remarkable" discovery, but still he stressed that it was for "plant breeders and for plant geneticists," not for widespread use in agriculture.[11] By changing combinations of darkness, temperature, and humidity, Lysenko had shown how it was possible to use tropical and subtropical varieties in breeding and genetic work to "an unprecedented extent."

Lysenko did not reply to either offer. Nikolai Ivanovich wrote of his disappointment to a senior colleague at the Institute. The congress was so important, he said, from all points of view in genetics—"theoretical, methodological and shifts in science are usually made at such gatherings. . . . It is not very pleasant to travel to the Congress on my own."[12]

It was much worse than Vavilov imagined. Lysenko would take advantage of his absence to plot his downfall.

CHAPTER 19

The Last Expedition

Nikolai Ivanovich launched his last foreign expedition, to Central and South America, in the late summer of 1932. After this tour he would never again be allowed to leave the Soviet Union. Now, at forty-five, he was as full of energy, curiosity, and resourcefulness as he was on his first journey into the Pamirs in 1916, but the opportunities for plant hunting and pursuing his scientific dreams were disappearing into Stalin's dark new world.

In the early 1930s, the party's Central Committee became the single patron of science. That meant any request for government funds for Vavilov's expeditions went straight to the top where Stalin's loyalists routinely consulted with state security before making a decision. Thus, at the beginning of 1932, when Nikolai Ivanovich applied for expenses for his next expedition to North and South America, the Central Committee sent his proposal to the OGPU's Economic Directorate for an expert evaluation. This was the same department that was in charge of building Vavilov's secret dossier, and its faceless servants advised, in a cold and inflexible tone, against the trip.

"For a number of years since 1924, [Vavilov's Institute] has sent numerous expeditions to different parts of the world, including America. It has gathered an international collection of seeds and plants. The collected material has still not been studied, and almost no practical conclusions and achievements have been introduced into the national economy—this work has never gone beyond the Institute's walls. The OGPU considers the organization of any botanical expedition to America inexpedient."[1]

Even though he eventually prevailed and was given the funds, Nikolai Ivanovich knew that it might be his last expedition. "I want to bring an immense amount of seed from America this time because it's unlikely I'll go again," he wrote Lenochka as he was sitting in his cabin aboard the steamship *Europa,* about to sail from the Latvian port of Riga.[2]

It was not only funds that were now hard to obtain. Getting a U.S. visa for the Ithaca congress had been surprisingly difficult. The American consul in Riga had mistaken Vavilov's membership of the Central Executive Committee, a national body of delegates with no political power, for membership in the Comintern, the Communist International whose agents carried out political missions abroad. Russian émigrés had written about his earlier expeditions suggesting that they were politically motivated, even that he was a Soviet agent trying to lure the best brains back to Russia, or steal their best ideas to take back with him. The U.S. consul in Riga had taken the reports seriously. Even though Vavilov had an official invitation to the congress and was to be its vice president, under Thomas Hunt Morgan, the consul interrogated him for an hour and a half. "I was talking sincerely and added to [the consul's] confusion—but it was clear he was sympathetic," he wrote Lenochka. "Still, I have visas to Peru and Bolivia [and] I also need Argentina and Brazil. That's the main issue. . . . Now India and China are the most important. And that will be it, the whole world."[3]

In fact, getting visas to South America had not been so easy. The consuls demanded documentation showing that he was not affiliated with anarchists. In Chile, he would be arrested, briefly, and his lecture about Soviet agriculture at the university in Santiago would be canceled at the last minute because the university authorities were frightened by the enormous crowd of people planning to attend. In Mexico, his return brought more stories in the local press about "plundering Bolsehviks." Even so, he managed to give nine lectures in America, not counting those in Ithaca. He gave a talk in Brazil and wrote an article for a Chilean newspaper—all promoting his idea to use the world's genetic diversity to improve food crops.

For all the concern in Moscow, Vavilov, once again, performed admirably as an ambassador of Soviet science. The Ithaca congress was a spectacular event attended by all the top geneticists of the day. Thomas Hunt Morgan and Hermann Muller were there, along with the researchers from Morgan's fruit fly lab. Richard Goldschmidt and several young scientists from Germany arrived, as did J. B. S. Haldane, Cyril Darlington, and R. A. Fisher from England, Roger de Vilmorin from France, and others from Denmark, Belgium, Switzerland, Spain, Italy, Canada, and Poland. Vavilov's friend Dr. S. C. Harland, the cotton specialist from Trinidad, also attended. There were four hundred booths and six hundred microscopes for looking at slides.

Vavilov reported that his Institute and its stations across Russia were experimenting with "300 different cultivated species."[4] From his expeditions he had collected "not less than 300,000 specimens," and "little Abyssinia has proved to contain more than half of the varietal diversity of cultivated wheat found in the world."[5] Vavilov continued to promote his Law of Homologous Series, but it was a more cautious version than the one from the heady days of Saratov. "The regular repetition of series of characteristics in varieties in different allied species and genera, is quite general," he said. "Every month brings new evidence of this astonishing parallelism which cannot be neglected in the study of the evolutionary process." But he now admitted that the parallelism he had seen was in the phenotype—in the observable characteristics of the plants, not in the types of genes carried by them—and therefore this parallelism "should not be taken as something absolute."[6]

He praised Lysenko's vernalization methods for "transforming winter varieties into spring ones and late varieties into early ones." But, importantly, he did not suggest that the work could be applied to large-scale farming. Rather, it opened "enormous possibilities to plant breeders and plant geneticists of mastering individual variation."[7] For his own grand scheme, he urged continued strong international cooperation and "the removal of barriers impeding research" of the "vast resources of wild species" of cultivated plants. "We are only at the start of our work in this direction," he concluded. "There is enough work for all of us."[8]

After the conference he flew "in a hurry" as usual first north to Canada and then south to Mexico and South America. In Canada, he collected fast-ripening varieties of wheat with rust resistance that would be of "exceptional" interest to Russia. In South America, he arrived in the spring when the potato plants were in flower and "when it is particularly easy to distinguish varieties." He became convinced that the many local varieties could be split into hundreds of different forms. "Our ignorance concerning the Andean potato diversity is striking. . . . There is a damned multitude of wild species and the cultivated potato is such as I have never seen before," he reported back to his Institute.[9] "I am taking everything that I can. It will prove useful. The Soviet Union needs everything. It must know everything in order to put itself and the world on the road!"[10]

Once free of the repressive atmosphere in Moscow and Leningrad, Nikolai Ivanovich regained his old optimism. From afar, he wrote, the Institute's work looked "even clearer. . . . We are stirring up the world. And we will get through to the heart of the matter.

The Institute's work is of great national and worldwide importance . . . we will continue the revolution that we have started in the plant industry." The Soviet Union was now leading the world, he wrote. "The U.S. Department of Agriculture has even followed in our footsteps in search of material for breeding potatoes . . . the same was done by the German Department of Agriculture."[11]

Vavilov sent six parcels of potatoes—at a hefty eleven pounds each, from Peru. "I cannot help sending them," he wrote, clearly overjoyed to be back on the road of discovery. "Tomorrow I will be at the Bolivian border . . . in the Cordilleras . . . in a month I will study the future of the world's agriculture in Argentina and Brazil. I am in a hurry."[12]

In Bolivia, he collected seeds of the cinchona tree, which was of great strategic interest to Moscow. Quinine used for the treatment of malaria is extracted from its bark. Vavilov traveled far from the cities and even small towns to the rough eastern slopes of the Andes to find these seeds. The seeds that he sent home would eventually become quinine-producing cinchona plantations on the Black Sea coast.

In Argentina, he collected "a complete set of grain crops improved by breeding, together with the best varieties of flax, corn and wheat bread." In this case, the cargo he sent home was not a few extra pounds. He shipped almost seven tons of material back to Russia.[13]

Brazil was the climax of the trip—his last few days as an explorer. He wrote, "Before me was a colossal country which occupies three-quarters of the continent of South America, a country in size exceeding Australia and North America and equal to two fifths of all Europe."[14] He flew in a light plane over prairie, forest, and jungle. He drove into the thick forests of the Mato Grosso, and he took a boat up the Amazon where it "rains every day and several times a day." He cheekily advised his colleagues to take an umbrella into the rain forest, a good joke and a good idea as well. He was impressed by the Biological Institute in São Paulo, which had preserved a tract of a typical rain forest. He wrote about the cicadas, the hummingbirds, and the "amazing large" butterflies, birds, parrots, and sometimes even the roar of a jaguar. "Every naturalist should visit the tropics at least once to experience all the violent development of life, the whole range of colors of the animal and plant world, all the complex transitions from life to death, from epiphytes to parasites to the creative force of life."[15]

As he moved from one exotic vegetation to another, Vavilov gave his own travelogue, complete with menus. When he finally saw the rubber plantations of Pernambuco on Brazil's east coast, he described rubber trees so large it took four men to reach around them. At one feast deep in the Amazon, he described the fare— boiled alligator, "like fish jelly with gristle," then fried yellow monkeys, "which had a peculiar and not very pleasant taste," red snakes, "not unlike sausages but more compact," and, finally, fried parrot, "which had a rather pleasant flavor."[16]

During all the adventures in South America, during his experiences of the comfortable living of his American colleagues, Nikolai Ivanovich never showed any sign of being tempted to abandon his mission in the increasingly difficult and dangerous Soviet Union. The hundreds of microscopes at the Ithaca congress, the beautiful campuses of Cornell or the University of Chicago or the University of California, had tempted others, notably Theodosius Dobzhanksy, but Vavilov apparently never coveted the American dream. When he saw the Japanese single-mindedness at work in Brazil, or the Ford Motor Company's "tireless work and typical enterprising Yankee energy"[17] extracting rubber on the banks of the Amazon, Vavilov knew that the Soviet Union in its present state was incapable of such enterprise. Yet he still talked about the promise of communism, how Russia would do better for its people, and not for one company bound by the profit motive, but for the entire world.

He was not a communist, but he was a socialist and a humanitarian. Whatever the apparent advantages of America, Vavilov truly believed that the opportunity to practice his beloved science would be greater in Russia than in the capitalist West. In any case, he would never abandon his family—his son Oleg, his beloved Lenochka, and now his second son, Yuri, who was born in 1928.[18]

While Vavilov must have assumed that he, like other bourgeois scientists, was under constant surveillance by the secret police, even when he was abroad, he did not know that Stalin had recently expanded the OGPU, increasing the emphasis on counterrevolutionary activities. Even as he was bringing priceless botanical treasures back to his Leningrad seed bank, Stalin's agents abroad kept him under surveillance, building a case that on his foreign expeditions he acted as an anti-Soviet agent. They were especially interested in Vavilov's two stops on his way home to Russia—one in Paris and another in Berlin.

In Paris, as was his custom, he called on friends and colleagues. Those included researchers at the Vilmorin seed firm, and a Russian émigré biologist, an immunologist at the Pasteur Institute named Sergei Metalnikov. He was eleven years older and had met Vavilov when he was a professor of zoology in Leningrad before the revolution. After the Bolshevik takeover, he was sent to the Crimea to organize the Crimean University, but was cut off from Soviet Russia by the White forces and left for France. He was on file in OGPU headquarters as a member of a group named Torgprom (Trade and Industry), which, according to the file, was "financed by American capitalist circles and the moving force behind the counter-revolutionary organization of veterinarians and organizers of bacteriological warfare against the USSR."[19] At a Paris railroad station Vavilov was spotted by an OGPU agent embracing Metalnikov on the platform. The agent immediately informed Moscow of this subversive encounter.

Vavilov also gave an interview to the Paris newspaper *Paris-Midi*, in which he talked up the achievements of Soviet agriculture, but the OGPU agent still found something sinister. Vavilov met the reporter at his hotel, and the result was a charming, innocent report.

"He stands up as I enter. He speaks French with ease, almost without an accent," the article began.[20]

Vavilov said, "I am extraordinarily busy. Life is short. I must call at the Pasteur Institute. We'll take a taxi and you will ask me questions along the way."

The French reporter was infatuated by this man of science. "There are people blessed by the Gods, whose intellect illuminates their whole face. They enter life without help of any bureaucracy; these mighty men of intellectual work. Here is one of them, no doubt about it."

Asked about his work before the revolution, as an employee of the imperial government, Vavilov rebuked the reporter.

"What's this question about? In Europe you always talk about a government. In Russia, even in tsarist time, we spoke about the state. By 1916 I already was a state employee . . . [after the revolution] I stayed on as the state's servant, an employee of the Russian state, my motherland's state. This is only natural."

Vavilov explained that he had been on an expedition for six months—one of many trips to fifty countries "to their most inaccessible places." In South America, he had just crossed Bolivian and Peruvian mountains "on a mule's back."

"Why?" the reporter inquired.

"I am extremely ambitious," Vavilov replied. "I am interested in global agrarian philosophy. [Excuse me] I should have said 'Marxist agrarian philosophy.'"

The reporter asked him to be more precise.

"I will quote Marx to you, 'Before scientists used to study the world to understand it; we study it in order to change it.'"

"To improve it?"

"This goes without saying."

"How are you going to do that?"

Vavilov then told the reporter about crossbreeding and about his world collection—"an international method . . . [of improving agriculture. . . . You can consider it utopia today. It's the future. It's a hundred year plan."

"A hundred years?" the reporter exclaimed, astonished at the time frame.

"It will take a hundred years," Vavilov replied calmly.

At this point, the taxi arrived at the Pasteur Institute and Vavilov took his leave. "Apologies, monsieur, life is short, time has no patience."

When the OGPU agents digested the article they distorted it just enough to make it look as though Vavilov was criticizing the Soviet leadership, and communism. In the OGPU dossier, Vavilov was quoted as saying, "I serve my country, not the government. I was formerly a tsarist assistant professor and stayed on in my country which is the same Russia as before."[21] The implication, of course, was that communism was a temporary aberration and the status quo ante would soon return. The agent had done his job well.

In Berlin, Vavilov visited a branch of the Kaiser Wilhelm Institute where Hermann Muller, the American geneticist, was working.[22] In Germany, Muller could not have arrived at a worse time. Hitler had just been appointed chancellor in January 1933 promoting his Eugenic Sterilization Act, and would soon be authorizing the arrest and harassment of Jews. Muller was partly Jewish, and a socialist. This was one reason why Vavilov offered him a position as director of genetics in his Leningrad Institute. Muller readily accepted. The offer, noted Muller's biographer, Elof Axel Carlson, was an "opportunity to do research Muller had only dreamed he might do. Full-time research, a generous budget, a large graduate program, the backing of Vavilov, one of the most powerful men of science in the U.S.S.R—all of this could not be passed up."[23]

For his part, Vavilov considered Muller's acceptance a coup for Soviet science. He had convinced one of America's best genetics researchers to come to Russia. But for the OGPU this was not a generous, patriotic act; it was yet another example of Vavilov's pandering to Western science and fraternizing with Americans, another stone to be used against him.

CHAPTER 20

Thunder and Dragons

In the winter of 1932–33, the Soviet Union was in the grip of another famine. Five million people, mostly in the agriculturally rich Ukraine, would perish. The famine was a direct result of the forced confiscation of grain from the farms to feed the cities. Instead of issuing an international appeal, as Lenin had done in 1921, Stalin concealed the disaster even from his own citizens, and found a scapegoat.

Late one night in March 1933, after Vavilov had returned from America, a colleague from his days at the Petrovka academy arrived unexpectedly at his Leningrad apartment. Lidya Breslavets was unusually flustered. She had heard that Vavilov was to be summoned the next day to the Kremlin where he would be reprimanded for wasting time—and money—on his foreign trips. But Vavilov did his best to calm her down. He was sure everything would be all right, he told her. His last trip to the Americas had been very successful. He had brought back thousands of seeds and samples, seven tons from Argentina alone, plus seeds of the cinchona tree. "There's nothing to worry about," he reassured Breslavets. "The people in the Central Committee are not stupid; they'll understand."[1]

But Breslavets's information was correct. Vavilov was summoned to appear before the Council of People's Commissars, the body that controlled state spending. A number of the Institute's former employees had complained about his expeditions, and more generally about his running of agriculture research. Essentially, the council wanted to spread the work of the academy away from Leningrad and out to the branch institutes. Frustrated, Vavilov wrote to a colleague, "Surely the academy is not coping with its job . . . it lacks strong financial management, there are no agencies to take care of supplies and not even a scientific secretary. . . . I am literally squashed under my numerous obligations."[2]

Once again, Vavilov explained to the commissars that the financial and the administrative burden had become overwhelming and he offered to resign, but again his offer was not accepted. Vavilov remained in charge of the academy for the time being, but he became a *nevyezdnoy*, a term for a person who was never allowed to travel abroad again, even to an international congress.

In addition, the OGPU had stepped up its recruitment of informers inside Vavilov's Institute. One of Vavilov's researchers recalled, "Starting in 1931–2, the 'Bolshoi Dom'—the headquarters of the OGPU in Leningrad—invited young employees—lab assistants—of the Institute to become informers. I was called three times and they dragged me through the conveyor belt. But I refused to co-operate. They would tell me, 'In your Institute, according to our information, there are wreckers, and you refuse to report anything even though you are from the proletariat.'"[3]

Within a month the secret police began to arrest members of Vavilov's staff, including some in senior posts. "Alarming events are piling up, twenty people have disappeared," Vavilov wrote to his friend Andrei Sapegin, Lysenko's boss in Odessa.[4] Among them were Viktor Pisarev, the director of the Pushkin experimental station outside Leningrad, Grigori Levitsky, the head of the cytology department (later set free), and Nikolai Maksimov, the Institute's leading plant physiologist. Maksimov had been the first scientist to criticize Lysenko's vernalization at the 1929 congress, pointing out correctly that it was not a discovery.

The arrests, a staff researcher at the Institute would recall, came suddenly. "Out of the blue, there would be a thunder clap, a dragon would roar and grab one of us. It was so scary. No one knew where the 'dragon' came from and why he picked on one member of staff and not another."[5]

After one of the arrests, Nikolai Ivanovich was driving home with a colleague from a meeting at the Academy of Sciences in Moscow.

"What is happening?" the colleague recalled Vavilov asking. "It's totally beyond any understanding . . . it's either a denunciation [a charge of political incorrectness] or a mistake . . . maybe they'll soon arrest us all."[6]

"No they won't arrest us . . . they have slandered us, but we are still alive," his colleague replied, trying to reassure him.

Despite the arbitrary nature of the arrests, Vavilov's staff naively believed that the police would behave according to the rule of law. To be a plant breeder was such an innocent occupation that if one was arrested there had to be a good reason. The accused had to have done something wrong.

"But it cannot be that Mikhail Grigorievich [the most recent researcher arrested] was guilty of anything," his colleague continued.

"Of course not," said Vavilov.

Soviet state security had never needed a legitimate reason for an arrest. In some cases, the charge was past membership of the fictitious Peasant Labor Party, which had been "destroyed" by the OGPU in 1930. In other cases, the arrested researchers were accused of various fabricated crimes of "wrecking" and "sabotage." Often, those arrested were asked, under severe interrogation, to provide testimony against their boss, Nikolai Ivanovich. Some refused, others could not withstand the pressure and signed "confessions." Some were found "guilty" and exiled to the north, or Siberia, others came back to Leningrad and told their horror stories.

The arrested researcher Mikhail Grigorievich was one of those released. He told Vavilov that his interrogators had tried to make him sign a false confession, but he had refused. Another researcher, the head of the department of forage plants, told Vavilov that he had been interrogated around the clock and when he couldn't stand it any longer, he had made false accusations against other members of the Institute.[7] He was set free and considered it "his duty" to tell Nikolai Ivanovich.

"I didn't blame him," Vavilov told a colleague, "I felt great sorrow for him . . . and yet some contempt."

In fact, Vavilov courageously complained about any arrested staff members to his boss, Yakovlev, at the Agriculture Commissariat. In those days, coming to the defense of an arrested colleague was a crime against the state.[8] His police dossier would record that Vavilov did everything possible to free the victims. "He took petitions from the accused and their wives and wrote appeals stating the innocence of the arrested and handed a list of the forty-four persons who he considered should be released to Yakovlev."[9]

And he instructed others on his staff to boost morale at the experimental stations where members had been arrested. When three researchers were arrested in Sukhumi, on Georgia's Black Sea coast, he instructed one of his senior staff to "cheer everyone up . . . help them in a comradely way, and ask the locals to support the honor of the Institute."[10] In this case, the three workers were released.

The Institute was now under attack from two directions: openly on agricultural theory and production from Yakovlev and Lysenko, and covertly from the police. The OGPU was constantly adding to Vavilov's "operational file"—a file prior to arrest. The file was kept at the OGPU's Economic Directorate, whose agents informed Stalin personally of its contents.

A letter to Stalin, undated but thought to have been written at the end of 1933 or the beginning of 1934, from the head of the OGPU's Economic Directorate, made a powerful case for Vavilov's immediate arrest. It accused Vavilov of membership in an anti-Soviet organization, of having anti-Soviet political views, of engaging in "wrecking," unapproved contacts with "foreign and émigré groups," and undesirable connections to high officials in the French government.[11]

As this darkness was descending on Vavilov and his science, there was one encouraging moment. In the summer of 1933, Hermann Muller arrived to work at the Institute of Genetics in Moscow, then part of the Academy of Sciences and under Vavilov's control. After his failure to persuade Dobzhansky to come home, it must have been heartening for Nikolai Ivanovich to be able to welcome his friend from Texas.[12]

Most cheering was Muller's arrival at the train station. He brought more lab equipment and personal luggage than a Soviet scientist could possibly imagine in those days. To ensure he could start work immediately on his fruit flies, Muller had brought ten thousand glass vials, one thousand bottles, two microscopes, and a suitcase packed with vials of flies and food-making equipment for them. In addition, he had shipped his 1932 eight-cylinder Ford, two bicycles, and trunks of clothing, books, scientific reprints, and household effects. He also brought his wife and his lab assistant, Carlos Offerman.[13] This was an example of the "Yankee can-do" approach Vavilov so much admired about America, and that was so out of reach to Soviet scientists.

Vavilov had arranged for Muller to be elected a corresponding member of the Academy of Sciences, but quite what the comrades thought of this American largesse one can only wonder. The Soviet press gave Muller a warm welcome. He was, after all, a well-known socialist who had been severely critical of American capitalism. In return, he toured Soviet cities giving speeches and wrote articles for the popular Soviet press, praising the friendliness of the people and the socialist innovation of the workers on the collective farms.[14]

Muller's arrival was not without risk for Vavilov personally, as well he knew. Muller had flirted with eugenics, and one of the Soviet ideologues' central arguments against Mendel and genetics was its use by the eugenics movement. In fact, Muller believed, like most geneticists, that social conditions, not a person's genetic makeup, had more to do with an individual's status and success in life and he vehemently opposed American sterilization policies and Hitler's own horrific persecution of Jews and minorities.[15] But he had proposed a controversial program that the sperm of gifted men be frozen and preserved as part of a voluntary eugenics program for future generations. This proposal would eventually get him into deep trouble when Stalin became aware of it—and it would make life more difficult for Vavilov.

For the time being, Muller's arrival bolstered Vavilov's international prestige among his fellow geneticists. Raissa Berg worked under Muller as a graduate student and she recalled, "I reported to Muller more dead than alive, overwrought at the prospect of seeing the great discoverer of the laws of nature."[16]

But such bright moments were few. Shortly after Muller's arrival, Vavilov wrote to the local Leningrad party committee of the desperate situation developing at the Pushkin experimental center where the *vydvizhentsy* had created a "hypertrophy of suspicion and mistrust" toward the specialists. Many were threatening to leave.[17]

He told the committee, "Briefly, comrades, the situation is not one of well-being, and I repeat once again that the collective and directorship must take a serious turn towards the creation of an atmosphere of trust, goodwill and sensitive comradeship, and not in words but in deeds so that it is felt in the quickest time before it makes us face even bigger difficulties." Again, the ideologues wanted to disperse the structure of Vavilov's Institute, to make the provincial experimental stations more autonomous, and again he successfully resisted it.

But the government then delivered an even more devastating and personal blow. In 1934, Vavilov planned to celebrate the tenth anniversary of his Leningrad Institute and the fortieth anniversary of the original Bureau of Applied Botany run by Vavilov's mentor Robert Regel. Nineteen thirty-four was also his twenty-fifth year as a scientist. As usual he fired up his staff to maximum effort and many guests, including several foreigners, were expected in Leningrad. The Institute received congratulatory telegrams from foreign agriculture ministries, including the United States, Turkey, Finland, Poland, and Bulgaria. Suddenly, four days before the ceremony was to take place, when the Institute was already decorated for the occasion, the Commissariat of Agriculture canceled the events.[18]

Vavilov was humiliated. He wrote a letter to the agriculture commissar, expressing his pain and indignation. The sudden cancellation had "depressed the whole staff, who feel as though they have received a vote of no confidence. . . . This naturally has prompted members of the staff to wonder whether they are fit to carry on."[19] There was no reply.

Vavilov was not about to give up, however. Typically, in the summer of 1934 he invited Lysenko to come to Leningrad and visit the Pushkin experimental farm. Vavilov even suggested that Lysenko should come "at least for a week two or three times a year to see what we are doing and to help, especially the young people, to faster and better implement the works of vernalization that we carry out here on rather a large scale."[20]

There is no record of a reply, but Lysenko did visit the Pushkin farm one summer when the flax was in bloom and it was possible to calculate the ratio of white to blue flowers and see if they matched the Mendelian ratio of three to one. One of the researchers recalled that Lysenko had "some sleazy guy with him and Nikolai Ivanovich said, 'So, here Trofim Denisovich, look and do your calculations, it's 3:1.' Lysenko looked at the figures, and said, 'Let me do the calculation myself.' Nikolai Ivanovich asked a helper to bring a bench for them to sit down. Lysenko sat down and got figures very close to 3:1. Then he stood up, looking unhappy and muttered, 'This is an exceptional case, and your Mendel and you raised it to the level of a general law.' Then, turning to his companion, he said, 'Let's go.' And that was the end of the encounter."[21]

Occasionally, Vavilov could not contain his frustration at such foolishness. One day at the Pushkin farm, he became very agitated about Lysenko's speculations. "Our magicians," he blurted, referring to Lysenko and his group, suddenly produce "unusual facts that were previously unknown—or were they simply unnoticed? What's the reason for this? Are we supposed to look for sudden changes in plants that cannot be explained by contemporary genetics? Our magicians do not take into account the years of long experience of world science, they do not want to hear the opinion of other researchers, they want to live in their birthday suits." Then he added, "Even a genius cannot figure out a complex issue if he is ignorant."[22]

Vavilov finally understood that his Institute was now under permanent siege, but he was still ill-prepared for the ferocity of the attacks on genetics and against him personally that would be launched by the hyperactive team of Lysenko and Prezent.

CHAPTER 21

The Lysenko Offensive

"Bravo, Comrade Lysenko, Bravo!"
—JOSEF STALIN, FEBRUARY 14, 1935[1]

When Stalin praised Lysenko in a Kremlin hall packed with plant breeders and collective farm workers, he marked a new and destructive era in the history of Soviet biology. A jubilant Lysenko quickly claimed a series of new plant breeding achievements, each one less credible than the last, yet each one carefully couched as "practical science," and linked expressly to Stalin's demands for quicker results.

The agricultural bosses followed their leader by giving Lysenkoism renewed support. Commissar Yakovlev praised Lysenko as a "practical worker whose vernalization of plants has opened a new chapter in agricultural science" . . . a scientist who was now "heeded by the entire agricultural world, not only here, but abroad as well." Lysenko's "people" would become the "backbone of the real Bolshevik apparatus."

Vavilov did not respond, at first. It was not in his nature to engage in an academic brawl—and, even if he had wanted to, the stakes were now much higher. For a geneticist to criticize Lysenko's claims when Soviet agricultural production was failing was to seem unpatriotic at best and, at worst, to be engaging in economic sabotage. Lysenko himself encouraged such treasonable thoughts. His speeches were routinely laced with references to the nefarious activities of "bourgeois scientists," "saboteurs," and "class enemies." In the Kremlin speech to the collective workers, the passage that drew the most applause, and Stalin's extraordinary exclamation, was not about any new achievement in crop production but Lysenko's attack on bourgeois science. All the bourgeois academics ever did was "observe and explain phenomena," he said, while socialist science aimed to "alter the plant and animal world in favor of the building of socialist society."[2]

Lysenko was a fiery speaker who knew his audience. "You see, comrades, saboteur-kulaks are found not only in your kolhoz life. . . . They are no less dangerous, no less accursed, in science. . . . Comrades, was there—and is there—really no class struggle on the vernalization front? . . . Indeed there was. . . . Instead of helping the collective farmers, they sabotaged things. Both within the scientific world and outside it, a class enemy is always an enemy, even if a scientist."

Isaak Prezent, Lysenko's viperish political minder, wrote the speeches, but Lysenko's delivery was masterful. He knew exactly how and when to grab Stalin's attention. It was the *kolhoz* system, the "mass of kolhoz workers," the collective farmers, who would provide solutions to Soviet agriculture, Lysenko said, not these "so-called scientists." And it was practical solutions like his own vernalization, solutions that came from ordinary, poorly educated farmers such as himself, that would give the collective workers a chance to prove their worth. Apologizing for his own lack of training, Lysenko had ended his Kremlin speech by emphasizing the distinction between himself and the academic theorists. He was not an academic, he was not an orator, he modestly declared, he was "only a vernalizer." It was at this point that Stalin jumped to his

feet and shouted his approbation and the Kremlin audience broke
out in wild applause.

Nikolai Ivanovich was in the Kremlin hall but remained silent, reluctant to confront Lysenko's faults—even when his close colleagues
pointed them out. At Vavilov's invitation, the British cotton breeder
S. C. Harland spent nearly four months in Russia. Nikolai Ivanovich
took him on a grand tour of the experimental stations, including
Odessa, and Harland would later recall, "I interviewed Lysenko for
nearly three hours. I found him completely ignorant of the elementary principles of genetics and plant physiology . . . and I can quite
honestly say that to talk to Lysenko was like trying to explain the dif-
ferential calculus to a man who did not know his twelve times table.
He was, in short, what I should call a biological circle-squarer."[3]

Harland found Vavilov "more tolerant" of Lysenko's short-
comings. Vavilov emphasized that the environment had never been
adequately studied by anybody, that the tenets of classical genetics needed to be revised to take account of environmental factors,
and that "young men like Lysenko who 'walked by faith and not by
sight' might even discover something. He might even discover how
to grow bananas in Moscow." Vavilov had said that "Lysenko was
an angry species," and that as all progress in the world had been
made by angry men, let Lysenko go on working. It did no harm and
it might do good.[4]

Some would criticize Vavilov for appeasing Lysenko; others
would see his stand as prudent. After all, there was evidence that
those who crossed Lysenko paid a heavy price. When Lysenko
arrived at the Odessa experimental station in 1932, the director
was a close colleague of Vavilov's named Andrei Sapegin. Sapegin
gave Lysenko his own laboratory and freedom to experiment with
his vernalization techniques, but insisted that all breeding be done
according to the Mendelian genetics. He required his staff to mas-

ter genetic theory and read the scientific literature. Lysenko simply refused. "It is better to know less, but to know just what is necessary for practical work today and in the immediate future," Lysenko used to say.[5]

Sapegin was naturally suspicious of Lysenko's claims. After the harvest, Sapegin noticed some sheaves of wheat on Lysenko's plot had been left un-threshed. He thought it was a mistake, or laziness, but then he discovered that the sheaves were on the control plots and had been left there on purpose. By not weighing the grain of the controls, Lysenko's staff had been able to exaggerate the relative yield of the experimental varieties. Sapegin was furious and made an example of Lysenko to the staff. Sometime later Sapegin was arrested as a "wrecker." He was imprisoned for two years and banished from Odessa, and Vavilov found a job for him in the Moscow Institute of Genetics.[6]

Vavilov had his own way of dealing with the "angry young" Lysenko. He recommended him for several scientific prizes and nominated him to be a corresponding member of the USSR Academy of Sciences. Vavilov's colleagues were aghast, but the party, which now controlled the academy, had already thrown its support behind Lysenko and fixed the vote in Lysenko's favor.

As Lysenko's star continued to rise, Vavilov's kept falling. The cancellation of Vavilov's celebrations for his Institute was just the beginning of the official disapproval of his work. The government now complained openly that Vavilov's Lenin Academy "had not fulfilled the basics task which it had been assigned."[7] After all the promises there was still a crisis in agriculture and in science.[8] But the most damning denunciation of Vavilov was made directly to Stalin himself.[9]

Two key party members, Alexander Bondarenko, recently appointed vice president of the Lenin Academy, and S. Klimov, the academy's party leader, wrote Stalin a secret letter. They considered it "their duty as Bolsheviks," they said, to let him know that since the cancellation of his celebrations Vavilov had become "particularly hostile." He "defended wreckers," and he was "always surrounded by most suspicious types." As president of the Lenin Academy he was now "a negative entity," only turning up for "festivities," and preferring to take foreigners on six-month tours of the Soviet Union rather than attend to academy business. He "longs to go to India, Persia and China"—"anywhere abroad" which seems to be evidence of his desire to be "far away from the USSR."

There was some truth in this harsh assessment. Vavilov did prefer expeditions to his administrative work, and he enjoyed taking foreign experts with him. He learned many useful things from them, and they would send him seeds for his world collection. But the authors of the letter did not stop there. They were appalled by Vavilov's reluctance to allow government inspectors into his Institute to check for "double-dealers, traitors, counter-revolutionaries and class-hostile elements." Vavilov insisted that his professors should be left alone to get on with their work, not subjected to prying officials. Vavilov "constantly and publicly announces that any inspection of the work of highly qualified scientists is simply insulting and 'unacceptable for him personally.'" This "resistance" was part of the class struggle. It was designed to protect the "old guard" and to keep the new party members, the *vydvizhentsy* who were "still weak and young in scientific terms," from making their presence felt.

Stalin made marks in the margin of the letter, underscoring the charges of "wrecking," "class struggle," and the concern about the lab inspections, and passed the letter to the Politburo. Within a few weeks, the Lenin Academy was reorganized, and Vavilov demoted to a vice president. In one sense he was happy to have less work to do, but he had lost a power base—the 111 research institutes and three hundred research stations were reduced to twelve main institutes. The government also tightened political control over the academy. The posts of president and academic secretary went to nonscientists, and of the fifty academicians, nine were appointed by the old Academy of Sciences and forty-one by the Commissariat of Agriculture.[10]

Again Lysenko seized the moment and in speeches and published works sharpened his criticism of genetics. For more than a year he had complained that geneticists had not sufficiently appreciated the role of the environment. Now, he rejected the core theory of genetics—that genes are passed on unchanged from one generation to the next—in favor of his own theory that the genetic makeup of an organism is subject to continuous change by the environment. In effect, he now supported the old nineteenth-century ideas of Lamarck and the inheritance of acquired characteristics, a politically acceptable move as Stalin himself had long been convinced of Lamarck's theory.[11]

In support of their neo-Lamarckist approach, Lysenko and Prezent published a booklet on plant breeding based on the theory that annual seed plants grow by stages, that the stages occur in sequence, and that special environmental conditions are required for each stage. This was one of Lysenko's better works, and the rational elements of it received widespread support.[12] But Lysenko went on to claim that by changing the development stages—by altering the temperature and exposure to light—it was possible to change the seed's genetic makeup and turn a late-ripening type into an early-ripening type and vice versa, in one generation. With the use of greenhouses to force the growth, the fourth generation of such plants could be reached within a year. All breeders had to do, according to Lysenko, was pick the offspring that ripened early and throw away the rest.

To geneticists who looked at the way dominant and recessive genes segregated when two plants were crossbred Lysenko's postulates made no sense. According to Mendel's laws, the situation was much more complex. Stable early-ripening types—a variety that could be relied on—would only appear after several generations. Still, Vavilov withheld direct criticism, making allowances for what he called the "psychological" situation in Odessa, for Lysenko's "nervousness." It was hard to find a way to confront youthful enthusiasm.[13]

But even Vavilov was running out of patience. Lysenko had become convinced that all varieties of plants deteriorated from year to year and that the way to halt this decline and reinvigorate the plants was for them to cross-pollinate—have sex, either within the same variety, or between two different varieties. This idea, as Vavilov would point out, was against all the tenets of traditional breeding. If a farmer had developed a good variety the last thing he wanted was to have its gene pool diluted by another variety through crossbreeding. Farms that grew seed grain deliberately kept their elite varieties separate. But without offering any evidence Lysenko simply declared his idea correct and all others invalid.[14]

Such demagoguery rejected outright the founding fathers of genetics—the English Bateson and the American Thomas Hunt Morgan—and Lysenko needed an authority for his new methods. He called up the "rich, scientific legacy of I. V. Michurin, one of the great geneticists," whose important work had been overlooked. Instead of "mastering this legacy," Lysenko complained, plant breeders were "interested only in how many foreign books someone has to read."[15]

The government sided with Lysenko. Vyacheslav Molotov, then chairman of the Council of People's Commissars and the formal head of the government, openly criticized Vavilov for squandering state funds on his expeditions. "Practical farming has not been made easier because this [world seed collection] is stored in the institute's cupboards." In contrast, Molotov praised the "brilliant" work of Lysenko. "Isn't it strange," he said in an editorial in the main agriculture journal, "that a number of scientists still have not found it essential to show Lysenko . . . active support?"[16]

Lysenko, at another meeting of farm workers, attacked scientists who worked "to no purpose" and "who argue about the incorrectness of [my] methods."

"Who? Why not name them?" asked Commissar Yakovlev.[17]

Lysenko named three of Vavilov's associates, Karpechenko, Tenis Lepin, and Anton Zhebrak. And then added Vavilov's name. "In the recently published work, Theoretical Principles of Wheat Breeding, Nikolai Ivanovich Vavilov, while agreeing with a number of theses advanced by us, also disagrees with our basic principle."

The next day Pravda announced that Lysenko had won the highest award of the Soviet Union, the Order of Lenin. There were now, officially, two separate doctrines in Soviet biology, Mendelian genetics and Lysenko's neo-Lamarckism. The battle for supremacy was engaged.

CHAPTER 22

The Showdown

At the end of 1936, the two camps in Soviet biology, one led by Vavilov and the other by Lysenko, presented their scientific arguments at the Lenin Academy in Moscow. Vavilov again tried to win time by arguing that the differences were really no more than normal disputes among scientists—plus a certain willful ignorance of genetics by Lysenko and his group. It might have been a good tactic five years earlier. Now it was too late.

From December 19 to 27, 1936, the Lenin Academy opened its doors to the contestants in Soviet biology. Stalin sent Nikolai Ivanovich and his fellow geneticists a chilling message to show where he stood—in case anyone was in doubt. Stalin's state security agents arrested a brilliant young geneticist named Israel Agol. In 1931, Agol had worked in Texas with Vavilov's friend Hermann Muller, and Muller was due to deliver one of four major speeches at the congress. Agol, a party member, was accused of deviating from the official line. For his perceived disloyalty, the young scientist would later be executed.[1]

The heavy political forces now acting against the geneticists were

underscored by the presence at the congress of the party's agricultural bosses—Yakov Yakovlev, now the Central Committee's
adviser on agriculture, Mikhail Chernov, the new commissar, and
Karl Bauman, the party's chief of science.

The congress was to evaluate the practical usefulness of the two
theories of plant breeding—Vavilov's and Lysenko's—so that the
bosses could pick the right one for Soviet agriculture. The idea that
an examination of the theories in two and a half weeks by deeply
divided partisans could provide the bosses with sufficient information to make such a choice was nonsensical, but that's how the Soviets were making farm policy in 1936.

Nikolai Ivanovich had devised a preemptive strike against
Lysenko and Prezent. In the foyer of the congress, he had asked
Muller to prepare a simple demonstration of the basis of genetics, including microscope slides of chromosomes in various stages
with color diagrams and written explanations. "Well, here at last
everything is perfectly clear," Vavilov declared. But the effort was
quickly dismissed by Lysenko and Prezent. They "bent over the
microscopes, [and] cast a cursory glance at the explanatory illustra-
tions . . . their entire examination took less than five minutes, and
they left the foyer without comment."[2]

Daily reports of the congress in *Pravda* built up tremendous
public interest and it had to be moved to a bigger hall to accommo-
date the audience that had expanded from seven hundred to more
than three thousand.[3]

The geneticists opened with an unprecedented barrage of criticism of Lysenko's methods and reporting.

An opening speaker asked Lysenko, "You cite harvest of tens of
millions of *poods* [1 pood = 36.1 pounds]. But where are the losses
that vernalization has caused?"[4]

Another speaker attacked Lysenko for not learning the principles
of modern genetics. Had he done so, the speaker argued, "Many of
the things he struggles to understand—like his 'marriage of love'

between two plants—he would find have already been explained
by genetics." Lysenko was reprimanded for his "shallow attitude"
toward science and his "unprincipled demagoguery."[5]

In his speech, Vavilov astonished his colleagues by continuing
his mild approach, playing down the differences between the two
sides. There was already sufficient agreement to produce an immediate practical solution to the problems of Soviet agriculture, he
said. His language was diplomatic and deliberately restrained, bely-
ing the furious battle in which the geneticists were literally fighting
for their survival.[6]

In contrast, Lysenko's tone was combative and uncompromis-
ing. There were indeed basic differences between the two sides, he
asserted, on three important issues: on Darwin, on inbred varieties,
and on changing the heredity of plants in a required direction by
"appropriate upbringing." In all cases, he singled out Vavilov as his
main opponent, as the chief representative of the geneticists who
were all following the wrong course. On Darwin the differences
were "impossible to reconcile." By citing mutations as the engine
of evolution, geneticists denied "a creative role" for natural selection in the evolutionary process. On his "marriage of love" idea,
Lysenko still maintained that forced crossbreeding within a variety
led to increased yield.

Lysenko also still insisted that plants could be "trained" to
change their inherited traits. He offered a single example: one plant
of a winter wheat forced over several generations by varying its
environment, changing its surrounding temperature and moisture,
to become a spring wheat. This proved that there were no immutable genes, as the geneticists claimed. Everything depended on the
environment.

For anyone with the most basic understanding of science, the
most astonishing part of Lysenko's presentation was his serious proposal that one experiment, of a single plant, could prove a theory.
As Zhores Medvedev, the Russian biochemist, historian, and chroni-

cler of Lysenkoism, would observe in 1969, "An experiment without replication is not a scientific experiment. . . . The single seed could have been a hybrid, a mutant, or a contaminant. . . . One casual seed does not represent a variety."[7]

But Lysenko didn't care what the academics thought. He chided Vavilov and the geneticists who he said "had let a good project slip from their hands." If only they would drop the idea of genes, he suggested, "they would easily come to the conclusion that winter plants at certain moments in their life, under certain conditions, can be transformed, can change their hereditary nature into a spring type, and vice versa. By adopting our point of view, N. I. Vavilov can also refashion winter plants into spring ones. Any winter variety can be made into a spring one."[8]

This declaration finally brought Vavilov to his feet.

"You can refashion heredity?" he challenged Lysenko.

"Yes, heredity," Lysenko shot back with no further explanation.

Lysenko would never be able to repeat his winter wheat experiment in any way that was scientifically convincing, and the idea would be dropped, just as his proposals for mass vernalization would eventually be shelved. But on the basis of this one wheat experiment, he now rejected entirely the chromosome theory, and denied the existence of the gene. Both were fabrications "invented by geneticists." Instead he proposed his own, new theory. "The hereditary basis does not lie in some self-reproducing substance . . . the hereditary basis is the cell, which develops and becomes an organism . . . there is not a single bit that is not subject to evolutionary development."[9] Thus, without any scientific papers, without any more experiments, without any more facts, Lysenkoism, a muddled mumbo-jumbo of neo-Lamarckism, was officially born.

For the political bosses, under pressure to deliver a practical solution, Lysenko had provided what appeared to them to be a legitimate alternative. The geneticists had failed to score the demolition points required to dismiss Lysenko. Some had even given a legitimate reason to doubt their own theory by leaving open the possibility that a strict interpretation of the immutable gene was indeed vulnerable; the role of the environment was not fully understood.

After Lysenko, one more speaker appeared for the geneticists, the American Hermann Muller. Vavilov had asked him to outline the canons of genetics, and he did this brilliantly, but Muller was a fighter and he could not let Lysenko's speech pass unchallenged—even if his friend and boss, Nikolai Ivanovich, had cautioned him to be moderate.[10] Using much tougher language than any of his colleagues, Muller branded Lysenko's ideas as "quackery," "astrology," and "alchemy." Muller also called attention to the fascist race and class implications of Lamarckism. If Lamarckism were true, he said, it would "imply the genetic inferiority, at present, of peoples and classes that had lived under conditions giving less opportunity for mental and physical development."

The effect of Muller's allegation would be devastating for Vavilov. The Kremlin had banned the topic of eugenics on the grounds that it could be used to promote racist policies—as Hitler was doing with his Eugenic Sterilization Act. Not only had Muller violated the ban, but he had accused the neo-Lamarckists, of whom Stalin was one, of pursuing a doctrine that could be used by fascists for racist ends.[11]

Early the next morning, Vavilov went to see Muller in his hotel room. Muller would recall that this was only the second time he had ever seen his friend "in obvious trepidation."[12] The other time was when Vavilov had accidentally bumped directly into Stalin one day as he was turning a corner in the corridor of the Kremlin. Stalin, ever paranoid about his own safety, saw that Vavilov was carrying a briefcase, stuffed full as usual, and may have thought, for a split second, that there was a bomb in it. He glared at Vavilov and darted into his office. Vavilov was badly shaken and it had left him, as Muller had seen, extremely unnerved for some time afterward.

On this morning in Moscow, Muller noticed Vavilov was in a similar agitated state. As related by Muller, Vavilov told him that linking the Lamarckist theory with fascism and racism had caused the managers of the congress "to debate heatedly through the night in the search of a policy to adopt with reference to me [Muller], to geneticists in general, and to the matter at issue. He pleaded with me to make a public retraction of some kind."[13]

Muller did make a statement, "absolving the Lysenko group," but he did not retract his view that "racism would be a necessary consequence of the erroneous doctrine of the inheritance of acquired characteristics."[14]

The damage had been done, as Muller would later admit. "Neither Vavilov, nor others, said any more to me about the matter, but there were evidences that the chasm between the opposed groups had permanently widened."[15] The consequences would be even more troublesome for Vavilov than he could have imagined.

As the congress ended and the press declared a triumph for Lysenko. The question was why Vavilov had not fought harder. Why had he not used scientific arguments to demolish Lysenko's anti-science speculations? Until finally challenged by Lysenko toward the end of the congress, Vavilov had hardly acted as the leader of a group of scientists determined to show their theory was superior. There was a reason. Vavilov did not want to emphasize a fight that would scare off his international constituency. He believed that he could win the genetics argument at home by playing the one card Lysenko and the opponents of genetics did not have, friends and admirers abroad. Vavilov had worked hard to develop this network, writing letters and inviting foreign researchers to Russia. They came, one after the other, from America, Britain, and Europe, and returned impressed with the scope of the Institute's work and with Vavilov personally. Many of them returned home telling of Vavilov's prodigious energy. In 1935, Otto Frankel, a young New Zealand plant breeder, came to Leningrad to meet Vavilov and examine the Institute's work on wheat and potatoes. Frankel recalled later, "When he dropped me at my hotel in the evening he usually had an armful of journals to look through during the night."[16]

Vavilov had made such an impression at the 1932 International Genetics Conference in Ithaca, New York, that Moscow had been chosen to host the next conference—in 1937. This was a great honor for Vavilov and for the Soviet Union. It was also an opportunity for Vavilov to demonstrate to his reluctant bosses, to other scientists, and to farmers, how genetics was being taken seriously and used successfully in plant breeding in other countries. Lysenko was as reluctant to have the conference in Moscow as Vavilov was keen. Such an international gathering would expose his shaky doctrines to even greater examination.

By the spring of 1936, the conference organizing committee, composed of representatives of several nations, had drawn up a draft program. They planned for 1,500 delegates, nine hundred Soviet and six hundred foreign, and "likely 200–300 family members." Seven hundred and fifty invitees had already confirmed they were coming. Vavilov was to be the president. By early summer, the Central Committee had still not approved the congress, but in November Stalin himself canceled the meeting. Stalin had written on the Central Committee's resolution, "People in charge of preparing the congress are useless, they came up with a proposal before they prepared the business. Will have to cancel."[17]

Disappointed members of the Academy of Sciences, including Vavilov, tried to turn the cancellation into a postponement and managed to persuade the Politburo to agree to hold the conference a year later, in 1938. But the problems had not gone away. Karl Bauman, the science adviser to the Central Committee, had raised one particular difficulty in an undated letter to Stalin and Molotov: Lysenko's adherence to the Lamarckian view of heredity.

Bauman wrote, "While the majority of geneticists in the U.S.S.R. and other countries hold that acquired characteristics are not inherited by the next generation, academician Lysenko, on the basis of works of Michurin and his own, insists [that] development of an organism [can result in] a change of inherited characteristics." While "all scientists recognize Lysenko's merits—his theory of stage development of plants and his methods of vernalization— many consider his overall genetic views to be wrong, contradicting modern science."

Bauman continued, "This creates a not completely healthy atmosphere . . . it [will be] necessary to provide a free discussion of the debatable questions of genetics, especially in connection with the upcoming international congress."[18]

Still hoping that the congress could be saved, Vavilov had wanted to keep the Lenin Academy meeting as low-key as possible, showing that the "controversies" could be overcome, but events turned against him.

On the eve of the meeting a report in *The New York Times* had said that a serious split had developed among Soviet geneticists and that some of them, including Vavilov, had been arrested.[19] In a furious cable to the newspaper, Vavilov complained about its "slanderous" and inaccurate report and of its "insinuations" that there is no intellectual freedom in the Soviet Union.[20] "Slander is the usual method of the Soviet Union's enemies; fascism is especially successful at it," he wrote. Referring to the suggestion of a bitter quarrel over genetics, he added, "We have arguments, we debate existing theories . . . we challenge each other . . . it is a strong stimulus that significantly raises the level of work." And he ended by declaring, "More than many others I am indebted to the government of the USSR for the enormous support given to the institute that I head. . . . As a true son of the Soviet land,

I believe it is my duty and pleasure to work for the benefit of my motherland, and to give myself to science in the U.S.S.R."

Vavilov's pleading did not convince the organizing committee of the international congress. They began looking for a new venue. They would eventually decide on Edinburgh, Scotland. Vavilov's last best chance to show his masters that genetics, not Lysenkoism, should prevail had slipped from his grasp.

CHAPTER 23

The Terror

The year 1937 marked the beginning of the second wave of arrests and show trials. Stalin demanded total loyalty to all decisions made at the top and those party members and military officers who failed the test were eliminated. Stalin would deliver his famous anti-Trotskyite speech on the need to liquidate "double-dealers." Nikolai Bukharin, the party's longtime theorist, was arrested and would be shot a year later. The Bolshevik old guard and the military were at the highest risk, but all critical discourse—social, scientific, and political—was now held within narrow bounds. In this terrible new repression, geneticists were branded as traitors, and the personal attacks on Nikolai Ivanovich increased.

At the height of Stalin's bloodletting in 1937, Nikolai Ivanovich's legendary optimism began to fail him. He wondered aloud to his family how much longer he could avoid arrest. As director of the Leningrad Institute and also of the Institute of Genetics in Moscow, Vavilov was constantly shuttling back and forth between the two cities. In Moscow, he always visited the apartment in Guzinsky Val where his mother now lived with the divorced Katya and their son, Oleg. On one of these visits, his brother, Sergei, was at the apartment. When the time came to leave, Nikolai Ivanovich started to speak "with unusual sadness" about the terrible times that he now saw coming.[1] When he took the train between Moscow and Leningrad, he said, he was never sure if he would reach his destination. He suggested to everyone present that they should be careful about what they say to others, especially about family members and politics. In these times there were always some who could not be trusted, he warned. Then he took a piece of paper and wrote these lines in English from a British government exhortation to its citizens during World War I:

> *If you your lips*
> *Will keep from slips*
> *Five things you must beware:*
> *Of whom you speak,*
> *To whom you speak,*
> *And how, and when, and where.*[2]

Katya begged Nikolai Ivanovich not to engage Lysenko. She told him to control himself and be more like his brother, Sergei, do his work and stay in the background. "If you keep demonstrating your pride, you'll perish," she warned him. "Hide that pride in your pocket. Proud people cannot live safely."[3] But it was too late for such advice.

Less than two weeks after the Lenin Academy congress, the party chief of agriculture, Yakovlev, and Lysenko's political henchman, Isaak Prezent, viciously attacked Vavilov in the press. And Alexander Kol, who had been informing on Vavilov for the state security for a decade, renewed his dirty work.

Yakovlev lashed out at the "bourgeois," "negative," and "ideal-istic" disciplines of genetics, branding geneticists as "reactionaries and saboteurs." He compared genetics to religion: the concept of the immortal gene and the immortal cell in which the genes were supposed to reside was the same idea, he said, as religion's "body and soul." He singled out Vavilov by name as the leader of this politically unacceptable science.

Still smarting from Muller's attempt to link Lamarckism with racism, Yakovlev turned the same charge on the geneticists. In some countries, he wrote, genetics "forms the basis for the theory of the superiority of a particular race, allegedly possessing the best genetic stock, or of the wealthy classes, ostensibly possessing a monopoly of especially precious genes." The Lysenko forces now proposed that the immutability of the gene, as put forward by Mendel, was incompatible with Darwin. Mendelian genetics, by suggesting genes could be passed unchanged from one generation to the next, denied progressive biological evolution and lent itself to the idea of pro-ducing a superior race. In this simplified construction, geneticists were anti-Darwin, and racist. For another scientific authority on this matter (besides the gardener Ivan Michurin), the Lysenkoists plucked out of the past Kliment Timiryazev, the one-time liberal professor at the Petrovka agricultural academy. Timiryazev was alleged to have said once that Mendelian genetics was a "clerical-ist and nationalist intrusion" into science, and that was enough to endear him to the Lysenko forces.

The idea that geneticists were anti-Darwin no longer made any sense. It was attacking genetics as it used to be in the early years, when geneticists drew a too emphatic distinction between the hered-itary genes and the rest of the organism, rather than as genetics was developing to include effects of the environment. But Mendelists would now be branded anti-Darwin, and attacks against Vavilov continued at a furious pace. Lysenko would now go into battle under the joint banner of Timiryazev, who had died in 1920, and the gardener Ivan Michurin, who had died in 1935, but the movement was also known as "Lysenkoism."

Yakovlev's attack also turned into a vicious public chastisement of Nikolai Koltsov, the eminent zoologist, cytologist, and experimental biologist who for two decades had run the Moscow Institute for Experimental Biology. Koltsov, who had been the chairman of the Russian Eugenics Society until it was disbanded in 1929, was accused of advocating fascist conceptions.[4] Yakovlev demanded that genetics now proceed according to Lysenko's theory of development "rather than be turned into the handmaiden of Goebbels' ministry."[5]

Yakovlev's onslaught was printed in Lysenko's journal, *Vernalization*. The next issue included a rant from Prezent accusing "the powers of darkness"[6] of disrupting Lysenko's work. Those "powers" included Trotskyites, the recently arrested Agol, the purged Bukharin, and also Vavilov, he argued. (Bukharin was arrested on February 25, 1937, and would be shot a year later as a spy and a saboteur. Bukharin had supported the geneticists against Lysenko, and Vavilov was guilty by association, according to Prezent.)

Alexander Kol, the secret police informer, joined in the barrage. He attacked Vavilov's expeditions. Vavilov had been "less interested" in selecting varieties useful for the Soviet Union than "in selecting morphological wonders to fill the empty spaces in his homologous tables," Kol wrote.[7]

The campaign against Vavilov and other "knights of the gene" was continued elsewhere in the press. Citizens were urged to be vigilant, and "give serious thought to the motives behind the Bukharin strategy."[8] Vavilov was attacked for "giving refuge" on his experimental stations "to men who cannot be politically trusted," including "a certain Sobolev, a former nobleman, banished from Leningrad. His assistant Gildenbrandt is a former landowner, also banished." The agriculture journal *Sotszemledelie* called for the Lenin Academy to be "sanitized" and for officials to "root out mercilessly enemies and their yes-men from scientific establishments."[9]

• • •

The heat was too much for some old Bolsheviks now in Stalin's sights. Nikolai Gorbunov, Lenin's former personal secretary, Vavilov's original patron on the Council of People's Commissars and now permanent secretary of the Academy of Sciences, withdrew his support in an effort to avoid his own arrest. He now accused Vavilov of failing to dissociate himself from the fascist theories of genetics. It would not be enough. Gorbunov would be arrested in September 1938 and shot. Four others, all party members and in high positions in agricultural research, were also arrested, handing Lysenko access to the top political and administrative positions in Soviet agriculture.[10]

And yet Vavilov remained free. Stalin apparently felt that he was already getting what he wanted—the rise of Lysenko—without having to deal with the messy problem of arresting a popular international figure. The *New York Times* episode showed how well connected Vavilov was, and how closely the world's scientific community was watching. In any case, Vavilov had been effectively neutralized. His international expeditions had been halted, importing of foreign publications had been curtailed, and he would no longer be recruiting foreigners to his Institute. In fact, he had already devised elaborate ways of advising potential recruits to stay away.

The American plant breeder Harry Harlan had been a close friend of Vavilov's for several years and his son Jack wanted to study in Leningrad. On his last trip to America in 1932, Vavilov had agreed to send Harlan a coded message if it became too dangerous for his son to come. In the spring of 1937, the son wrote Vavilov asking for a job and eagerly awaited a reply. But when it came, Vavilov's letter began, "My dear Dr. Harlan," and did not mention a job—only a description of Chinese wheat. This was a code; "My dear" instead of "Dear" and anything irrelevant like Chinese wheat meant don't come.[11] The elder Harlan read the letter aloud to his son and then declared, "Vavilov says you must not come. You could do no useful work [and] you would be in danger yourself."[12]

As Russia's geneticists disappeared and Lysenko's political clout grew, Vavilov became increasingly isolated. Even his old friend Muller had become disillusioned and decided to get out while he could. The question was how to leave without making it appear as though he was abandoning Vavilov and his colleagues, a move that was sure to be exploited by Lysenko and Prezent. Muller's socialist friends were offering their services to the Republicans opposing the forces of General Francisco Franco in the Spanish Civil War that had started in July 1936, and Muller decided that might be an acceptable excuse.

He found a job in Madrid testing a new form of blood transfusion that involved extracting blood from cadavers. He asked the Soviet authorities for "unpaid time off" for the trip to Spain. The ploy worked; he left for Madrid in March 1937 and kept up the pretense that the Spanish trip was only temporary.[13]

As it turned out, Muller was only in Spain for eight weeks during which he received occasional updates from Vavilov about Lysenko's activities and about the international genetics congress planned for Moscow. Vavilov described how he had visited Lysenko again in Odessa to examine his work. What he saw was even worse than he thought. "I must say that for us it was not much. I thought even more of the earlier work."[14]

Vavilov read about Muller's Spanish exploits in the newspapers. "From my point of view you have done well in going to Madrid. But be careful! Science needs you. We must still do much work. You made a splendid start and you must continue your work here, which is no less important than your work in Madrid."[15]

Vavilov was still hopeful that the international genetics congress could be held in Moscow. He told Muller, "Molotov [chairman of the Council of People's Commissars] and Litvinov [Soviet foreign commissar] are themselves now engaged in arranging that the Congress be held in the USSR. . . . So now we have started a campaign . . . and we ask you to help us in this matter. Of course, you understand that it is very important for the benefit of genetics to have it in the USSR."

Outside the USSR, Muller felt enormous pressure not to criticize Lysenko, fearing that to do so might cause further harm in Moscow and Leningrad. His reticence was increasingly difficult for his colleagues to understand. He wrote to Julian Huxley, the British evolutionary biologist, "It is hard for people living abroad to realize (1) how tight our mouths are kept closed where I have been, (2) that if anything is repeated in the outside world it soon comes back there and has an effect similar to what it would have had if made there, (3) that even though the isolated statements may in themselves be true, they are bound to give rise to misunderstandings, both ways, that overshadow the truth—especially in view of the lack of free, critical discussion. . . . It seems that, in a world of conspirators, almost any action one can take becomes a conspiracy in one way or another!"[16]

Having secured a temporary job in London at the Institute of Animal Genetics, Muller returned to Russia in September to pack up his things and found that two more members of the Institute of Genetics in Moscow had been arrested. Once back in Stalin's conspiratorial world, the American realized that now he would have to leave or die at Vavilov's side. "As time passed, the sad conclusion was forced on me that in the USSR genetics was passing under too much of a cloud for my return there to be of any help."[17]

On the night before Muller's departure, Vavilov arranged a small party for his American friend. Raissa Berg, the young geneticist who had worked with Muller, recalled the evening. "There was a small banquet. Then Vavilov took us to the movies to see *Peter the*

First. Later we walked around Leningrad. Muller spent his last night in Russia at the Hotel Angleterre. We parted at the entrance to the hotel. It was past midnight. As we all shook hands Vavilov said, 'We meet here tomorrow at five in the morning.'"

The following morning Nikolai Ivanovich brought apples as a treat. The three of them drove out to the experimental farm at Pushkin where Vavilov gave them a tour. Berg recalled that the harvest at the experimental farm that year was as bountiful as a farmer could imagine. "There was Abyssinian flax with its magnificent large blue flowers, heaps of pink potato tubers, resembling little piglets. In the cytology laboratory Grigory Levitsky showed us slides and in the bakery were little pieces of bread [made from experimental flour] that resembled Eucharist wafers. We were served tea, white bread, smoked fish and chocolate bars."

Berg was struck by the warmth and informality of it all. Vavilov's driver sat with them for breakfast. "My father for all his Tolystoyanism would not have seated a driver at the same table, nor would any other director have done so. In this regard, as in so many others, Vavilov was an exception."[18]

When they returned to Leningrad, Nikolai went to his next meeting and Berg saw Muller off at the train station. In Vavilov's study Muller had left an official letter explaining his departure. He had been about to close the door, but stopped and went back into the study. In the letter he had written that he would come back in two years, but he corrected it to say one year. Muller would be leaving behind a man he admired intensely, as a scientist, a friend, and a human being. "He [Vavilov] was a truly great man in very varied respects, scientific, administrative, human," Muller would later write. "He was also more life-loving, life-giving, and life-building than anyone else I have ever known."[19]

Muller did not come back in a year, or even two years. He moved west for good and for the scientific pursuits that were increasingly being denied his friend.

CHAPTER 24

Into the Pyre

Vavilov's mother, Alexandra Mikhailovna, died on April 5, 1938, severing the final link with the Vavilovs of Middle Presnya. At the same time, Lysenko became president of the Lenin Academy and effectively the boss of Soviet agricultural science. The public attacks against Vavilov turned into unrelenting persecution. Lysenko began undermining the Institute in Leningrad, forcing Nikolai Ivanovich to make a stand.

Nikolai Ivanovich was late for the funeral. Family and friends, about thirty altogether, had gathered at Alexandra Mikhailovna's graveside waiting in the cold of early April. As they wondered how much longer they should delay the service, they saw Nikolai Ivanovich half walk, half run across the snow, waving his hat, his heavy woolen overcoat unbuttoned and flapping. He was slightly lopsided, weighed down by his crocodile-skin briefcase stuffed, as always, with books and papers. He had been out of Moscow when his mother died and this was surely not how a dutiful son would have planned his last farewell. Still, she would not have been surprised. Nikolai Ivanovich had chosen to live his life in a hurry, his grand scientific visions taking priority over everything, including family. Even the funeral of his mother had to fit, somehow, into the larger scientific scheme.

Other members of the Vavilov extended family knew where they ranked in Nikolai Ivanovich's hectic life. He had two families, really, Yelena and Yuri in Leningrad, and the Presnya clan in Moscow—son Oleg and former wife Katya, his sister Alexandra and her family, and his brother Sergei and his wife and son. Each had learned to share Nikolai not only with his science, but with the other family in the other city.

Only the Moscow relatives stood by the graveside that April day. That was how Alexandra Mikhailovna would have wanted it. She had never approved of Nikolai Ivanovich's divorce from Katya, and had frowned on his romance with Yelena. That was surely one reason why Nikolai had made no special effort to bring his two sons together. Oleg was favored with trips—to London with Bukharin, to Kozlov to see Michurin, and to the experimental stations with Muller. But Oleg never got to know Yuri. They met only three times during Vavilov's lifetime—once when Nikolai Ivanovich took Oleg with him to visit the Institute's polar station on the Kola Peninsula. Yuri would sometimes write to Oleg asking him to buy things in Moscow, paints and books, but not much more.

Nikolai's brother, Sergei, might have brought the two families together, but he had always preferred Katya to Yelena. As Sergei waited for Nikolai to arrive at the cemetery he thought about the broken family his mother was leaving behind, so distant now from those days in the three houses in Middle Presnya. Sergei had been visiting Presnya's Vagan'kovskoye cemetery and the Vavilov family plot for forty years, ever since the death of babushka Domna, their maternal grandmother. He remembered her wake, the pie and honey in the old house by the cemetery. Then in 1905, the year of the uprising, came the funeral for Ilya, his younger brother, who died, aged seven, of appendicitis. Then their sister Lydia succumbed to smallpox in 1914. And now it was their mother's turn. The funerals had always been in the spring and Sergei would note in his diary that he had come to associate the powerful scent of hyacinths with death.

When Nikolai Ivanovich reached the coffin he stopped and burst into tears. As the priest read the service and the coffin was lowered into the grave, Nikolai Ivanovich started sobbing uncontrollably. After the wake Sergei and Nikolai walked for a while along the neglected paths of the cemetery, a piece of old Russia where many of the intelligentsia were buried. They spoke of Lysenko and the troubles, and their brief conversation ended with them both wondering how long either of them could last in Stalin's Soviet Union. Which one would bury the other.[1] Then Nikolai hurried off to another meeting and to complete a report for his new boss, Trofim Lysenko.

Lysenko would benefit handsomely from Stalin's terror. Two of the party bosses of agriculture had been arrested and would be shot, Chernov in March 1938, Yakovlev (despite his support of Lysenko) in July. Bauman, the Central Committee's science adviser and one of the most passionate supporters of collectivization, died in jail in October 1937, two days after he had been arrested. Alexander Muralov, who had replaced Vavilov as the president of the Lenin Academy, was also arrested and he would be shot in March.

In the spring of 1938, Lysenko was appointed president of the Lenin Academy, the top post in the hierarchy of agricultural science. He was now Vavilov's boss and was shameless in his efforts to undermine Vavilov's remaining power base—as director of the Leningrad Institute and the Institute of Genetics in Moscow.

At the same time, the NKVD, successor to the OGPU, increased its surveillance. By December 1938 Vavilov's "operating file" with its informers' reports filled more than five folders, and now the NKVD opened a new one: File No. 300669. Under the heading

"*Genetika*"—genetics—the first document in this file was a memo entitled "Struggle Waged by Reactionary Scientists Against Academician Lysenko." It called for Vavilov's immediate arrest and was signed by Bogdan Kobulov, deputy to Lavrenty Beria, head of the NKVD. The file included new reports from undercover agents who had been spying on Vavilov from inside the Institute.

One of the most active of these informers was Grigory Shlykov, the head of the Institute's Subtropical Department. In his official job, Shlykov wrote reports to the Central Committee and he had recently recommended that Vavilov be removed as director of the Institute. Shlykov reported that Vavilov was "constant and very artful in the suppression of initiative." Vavilov was also "a very gifted specialist, but muddle-headed in matters of theory and unquestionably not sincerely working for our [Soviet] system."[2] If he were removed the "Institute would cease to be a sinecure of one person."

But Shlykov was even harsher in his secret reports to the NKVD, calling Vavilov a double-dealer, a person who could not be trusted. Shlykov wrote that he was "convinced more and more" that Vavilov was a member of Commissar Yakovlev's group. Yakovlev's "outward negative attitude" had been "a cover for their actual relations as accomplices." The "baseness and cunning of these people . . . has no bounds," he warned his masters.

Shlykov also suggested that all the fuss in *The New York Times* at the end of 1936 about Vavilov being arrested was a ploy organized by Nikolai Ivanovich himself as self-protection. He concluded, "I am therefore turning to you and your entire system to take measures to expose these circumstances."[3]

The public attacks against Vavilov now turned into unyielding torment. An Academy of Sciences commission looking into his Institute of Genetics concluded that the Institute "not only does not combat hostile-class viewpoints on the biology front" but its mistakes "provide grist to the mills of antiscientific 'theories.'" The Institute refused "to acknowledge T. D. Lysenko's scientific works."[4]

In the Kremlin, Stalin issued orders to "fight the high priests of science" who had "retired into their shells" and to "smash old traditions, norms and viewpoints."[5] The Academy of Sciences joined in the criticism of Vavilov's Institute and by midsummer of 1939, Lysenko would be elected a full member of the academy. He used his new status to undermine Vavilov's leadership of the Leningrad Institute by appointing his own people to high positions.

One of Lysenko's appointments, a *vydvizhenets* named Stepan Shundenko, was made Vavilov's deputy. Shundenko was an informer secretly reporting to the NKVD. Some of the Institute's staff undoubtedly guessed the real loyalties of their boss. One recalled Shundenko as "sinister" and a "nervous figure with dark, penetrating and constantly darting eyes," and that he "quickly hit it off with another repulsive person, the graduate student Shlykov." The staff made up a jingle about the two men, describing Shundenko as a "puny little devil" and Shlykov as a "would-be Napoleon," and both of them made of the same substance—"shit."[6]

Whether Lysenko could identify which of his acolytes were working for state security is not known, but he was relentless in his efforts to undermine Vavilov's leadership, even trying to turn Vavilov's staunchest supporters against him. One day he invited Vavilov's young disciple Fatikh Bakhteyev to his office and lectured him for an hour about his erroneous position in genetics. He tried to persuade Bakhteyev to reject Mendelism in favor of his neo-Lamarckism. It didn't work. Bakhteyev was a scientist, not a sycophant.

For a time at least, there was a limit to Lysenko's new power; he could not remove Vavilov without the approval of the Central Committee.[7] But Vavilov himself could not be sure of the extent of Lysenko's new political support. It must have seemed to him that his career, and his life, were now hanging by a thread. He had been forced into a corner and was ready to make his stand.

He turned to his trusted lieutenant Georgy Karpechenko, at the Pushkin farm. "We've got to stop being passive and get active," Vavilov instructed his colleague in a letter.[8] "There's no choice. Only the most convincing facts—of which there are many by now—can be used to confront this obscene criticism." He urged Karpechenko to prepare reports immediately—"two or three articles must be written without delay." Wheat and potato were "the most convincing examples," he advised. "We have to prepare for the campaign smartly and not leave everything to the course of events. I think we can still win, if we are well prepared."

Vavilov would write his own spirited defense in the February 1939 issue of *Socialist Agriculture*. "To those who propose the elimination of modern genetics we say: first offer a substitute of equivalent value. Let chromosomal theory be replaced by a better theory, not a theory that sets us back seventy years."[9]

The same issue of the journal gave Lysenko the right of reply. The Soviet people would never be fooled by Mendelism, a "reactionary, idealistic, absurd falsification of science," he declared. The Soviet people possessed "an all-powerful weapon." They could read Stalin's *The Short Course of the History of the Communist Party* and they would understand that they could not "sympathize with metaphysics or Mendelism, because that is exactly what Mendelism is, veritable, undisguised metaphysics."[10]

Letters from geneticists who protested the journal's consistent support of Lysenko were not published, including one demanding, "Comrade editor, what aim is the newspaper pursuing in trying to deceive readers and discrediting itself and Soviet science in the eyes of our scientists, as well as scientists friendly to the USSR in capitalist countries?"[11]

In the middle of this ugly public fight, Vavilov kept writing to his colleagues abroad. Mindful, no doubt, that his letters were almost certainly being intercepted by state security, he pretended that his work continued normally. "Work goes well" at both the Leningrad Institute and the Institute of Genetics, he wrote to Muller. There were "very promising" results on quinine trees. He was "busy" with experiments at the Pushkin farm.[12] He was working on a new "ecological classification of cereals and flax," he wrote S. C. Harland. He was preparing his book on the evolution of cultivated plants.[13]

But in a later letter to Muller he admitted, "We are involved in a heated debate to defend Mendelism-Morganism. . . . The arguments that we had in 1936 have become even sharper. Most of the actors of this drama are the same people. . . . Some extremists from Odessa believe that Mendelism and chromosome theory have become outdated and must be replaced with . . . Michurin's and Lysenko's theory of development."[14]

In answer to queries about Lysenko from abroad Vavilov kept his distance and now referred to him, sarcastically, as "Dr. Lysenko."[15] Perhaps with an eye to his readers at the NKVD, he wrote to an American colleague who had inquired about Lyensko's work, "This issue interests us a lot . . . some of us are more optimistic, some eye it with pessimism. I can send Dr. Lysenko's work to you, titled 'Changing the Nature of Plants,' which is believed to be his major contribution."[16]

Despite Vavilov's efforts to continue as if nothing was happening, the staff was worried about his health. He looked exhausted—especially after a day in the lecture room at the Institute battling his critics. He came down from the podium "amid whistling and booing with his hair wet from perspiration, in a tone at once of restraint and bewilderment, but with indignation in his voice, having tried to explain sincerely his objections and to persuade his opponents that everything they were saying was the result of ignorance."[17] Some of his friends, seeing the strain, tried to send him on a holiday to a resort, probably on the Black Sea. Vavilov had hardly ever been on vacation; he would always protest that nothing should come between him and his science, not even a day off. For a long time, he resisted the offer and then agreed to go. Reservations were made for a trip to a resort—it was not clear if Lenochka and Yuri were included—but at the last minute he canceled. He did not have the time, he said.

As Vavilov entered his last full year of freedom, he finally realized that there was no longer any hope of reconciling the two sides. But he was still receiving foreign visitors. In the fall, Jack Hawkes, a young researcher at Cambridge University, came to study the Institute's potato collection and found Vavilov "very jovial and putting one at his ease with a most interesting conversation."[18] Vavilov told Hawkes that Lysenko was "not a real scientist at all and certainly not a geneticist," but he had "a marvellous knack of publicity and the government people are on his side." Hawkes understood that "the only thing" worrying Vavilov was that Lysenko was getting more money for his work.

After his visit, Hawkes sent a letter to *Pravda* speaking for a number of "English scientists" who, he said, used to admire Lysenko for his vernalization, but now "we find he takes no notice of genetics or reasoning . . . there is absolutely no evidence to show that the genetic constitution of a plant can be altered by the environment . . . What Lysenko is stating is merely another form of Lamarckism." The letter was never published.

Vavilov was now prepared to admit there was no retreat. At a meeting of regional scientific workers at the Leningrad Institute in March 1939, Vavilov clearly laid out the differences between himself and Lysenko. There were "two positions, that of the Odessa Institute and that of the Leningrad Institute," he said. "It should be noted that the [Leningrad] position is also that of contemporary world science, and was without doubt not developed by fascists, but by ordinary progressive toilers . . . and if we had here an audience of the most outstanding breeders, practical and theoretical, I am sure they would have voted with your obedient servant and not with the Odessa Institute. This is a fact, and to retreat from it simply because some occupying high posts desire it, is impossible.

"The situation is such that whatever foreign book you pick up, it goes contrary to the teachings of the Odessa Institute. Would you order that these books be burned? We shall not stand for this. . . . It is not to be solved by decree of even the Commissariat of Agriculture. We will go into the pyre, we shall burn, but we shall not retreat from our convictions."[19]

CHAPTER 25

Comrade Philosophers

The opportunity for scientific discourse was over. In the fall of 1939, the party's Central Committee organized a last forum between the two sides—but it would be run by Marxist philosophers, not scientists.

In his new powerful position as head of the Lenin Academy, Lysenko, the poorly educated vernalizer, the barefoot scientist from Kharkov, had become a monster, an arrogant, self-important party hack. He called Vavilov to Moscow to give an account of the Leningrad Institute, but he had no interest in the constructive scientific dialogue Vavilov had once offered him. He had no interest in Vavilov's report of his international seed collection, now including 250,000 samples and the biggest in the world. He was only interested in Vavilov's humiliation as a scientist and in his total defeat. For this odious exercise Lysenko employed one of his stooges at the Lenin Academy, a schoolyard bully named Lukyanenko. The May 1939 session was recorded by a stenographer and here is a sample of the interrogation.[1]

LUKYANENKO: You consider that the center of origin of man was in one place, and that we [Russians] are on the periphery?

VAVILOV: What is doubtless the case is that mankind originated in the Old World when there were no men in the New . . .

L: I don't believe that man originated in one place . . .

V: I have told you, not in one place but in the Old World . . .

L: This is connected with your views on domesticated plants?

V: My basic idea is that the same species does not arise independently in a variety of places, but spreads through the continents from one identifiable region.

L: Everybody says that the potato came from America. I don't believe it. Do you know what Lenin said?

V: The facts and historical documents attest to it . . . it is well known that potatoes were brought to Russia by Peter the First.

L: How do we know?

V: There are precise historical documents. I could with great pleasure tell you about it in greater detail.

LYSENKO (interrupting): Potatoes were brought into Russia. This is a fact. One cannot go against facts. . . . The question is whether, if the potato originated in America, it means that in Moscow, Kiev, or Kharkov it could not arise from an ancestral species until the Second Coming? Can new varieties arise in Moscow, Leningrad, or any place? I think they can. And then how does one view your theory of the centers of origin, that's the point.

LYSENKO continued: I understood from what you wrote that you came to agree with your teacher Bateson that evolution must be viewed as a process of simplification. Yet in Chapter 4 of the *History of the Party* it says evolution is an increase in complexity . . .

V: When I studied with Bateson . . .

LUKYANENKO: An anti-Darwinist.

V: No. Some day I'll tell you about Bateson, a most fascinating, most interesting man.

LUKYANENKO: Couldn't you learn from Marx? . . . . Marxism is the only science. Darwinism is only a part, the real theory of the knowledge of the world was given by Marx, Engels, and Lenin.

V: I studied Marx four or five times and am prepared to go on . . .

At this point Vavilov tried to end the confrontation. He remarked how difficult communication had become between the two sides, how their arguments had gotten in the way of any mutual understanding.

V: . . . Unfortunately our language has become clumsy, and specialized . . . difficult to understand, not only for other specialists but even for botanists. . . . We do not understand each other, yet we discourse of great things. We have worked methods of studying plant life, but to understand each other we must first learn the vocabulary . . . the one sidedness of which you talk is a deep untruth . . . subtle games are played. . . . You can imagine how difficult and complex it is to guide graduate students when all the time one is told that one does not share Lysenko's views. History will indicate which of us is right.

Then Nikolai Ivanovich admitted:

I am an overburdened man; not only do I work as the academic secretary, the deputy, but even as a financial administrative assistant . . . I should have explained this is greater detail. . . . If Trofim Denisovich would only listen calmly instead of shuffling pages . . .

LYSENKO: I agree with you Nikolai Ivanovich. . . . It is somewhat difficult for you to carry on your work. We talked of this many times and I was sincerely sorry for you. But you see, you are being insubordinate to me. . . . I say now that some kind of measures must be taken. We cannot go on this way. . . . We shall have to depend on others, take another line, a line of administrative subordination.

The meeting passed an unprecedented resolution, proposed by Lysenko. It declared the work of Vavilov's Institute unsatisfactory, and Nikolai Ivanovich would soon learn the meaning of Lysenko's chilling conclusion that "some kind of measures" had to be taken. Vavilov would be accused of a "systematic campaign to discredit Lysenko as a scientist"—a crime against the state, according to the minister of the interior, Lavrenty Beria, the head of the NKVD. Beria asked Molotov, chairman of the Council of People's Commissars, the highest governmental body, for his consent to arrest Vavilov, but again there was no response. Evidently, Stalin still did not feel that the time was right.

In the summer of 1939, Vavilov's colleagues noticed how much he was suffering. "He became somehow less colorful in that last year," wrote one his colleagues. "The old sparkle went out of his eyes and he lost his usual slightly ironic cheerfulness." The doorman at the Institute in Leningrad noticed how Vavilov was short of breath after climbing up the grand staircase. "It's my heart, my dear fellow," Vavilov confided.[2]

A group of Vavilov's colleagues bravely complained to Molotov that the attacks were destroying a great scientist. Molotov sent a team of inspectors to interview Vavilov, but they arrived just as he had received an envelope from abroad containing some new barley seeds. Thrilled, as always, to receive new additions to his world collection, he told Molotov's delegation that everything was "excellent" and offered them a look at his new seeds. When the inspectors suggested that he might be under strain from the personal attacks, Vavilov dismissed them as "trifles." Molotov's team reported back that Vavilov was fine and that his colleagues had raised a false alarm.[3]

In October 1939, the Central Committee, now eager to close the debate, arranged one last meeting of the two sides, but the organizers of the forum were Marxist philosophers, not scientists, ensuring that ideology, not science, would dominate the discussion.[4] There were sharp exchanges between Vavilov and Lysenko.

Vavilov had clearly reached the limit of his endurance. There was nothing to be gained any longer in avoiding a blunt confrontation. Lysenko's position ran "counter to all of modern biology," Vavilov declared. "In the guise of advancing science, we are advised to essentially turn back to scientifically obsolete views from the early to mid-nineteenth century. . . . What we are defending is tremendously creative work, with precise experiments by Soviet and foreign scientists as well."[5]

Vavilov also spoke about the intolerable drift to isolation in Soviet science. He warned about the Soviets being dismissed by the rest of the scientific world and being left behind, for example, by the Americans with their hybrid corn that was producing such impressive increases in yield. For a man who had spent his entire life looking at science from a global perspective, the isolation could not be more destructive, both personally and scientifically.

In response, Lysenko declared, "I do not recognise Mendelism . . . I do not consider formal Mendelism-Morganist genetics a science . . . We Michurinists object . . . to rubbish and falsehood in science."[6]

The chairman of the forum, a Central Committee–appointed philosopher named Mark Mitin, summed up the debate: Lysenkoism was a progressive science and genetics was reactionary. The philosophers had spoken. Vavilov wrote immediately to Mitin, deploring the outcome.[7] "The conclusion you drew at the conference on questions of genetics left us with a bitter aftertaste," he wrote. Although Mitin had noted the importance of Mendel's law and Morgan's chromosome theory, Vavilov said he was completely mistaken to divide genetics into "progressive" (Lysenkoist) and "reactionary" camps. Lysenko's speculations were fake, he charged, adding that Mitin's summing up could help perpetuate Lysenko's falsehoods.

The proceedings of the forum were written up in the party's journal *Under the Banner of Marxism*, giving Lysenko's side the party's seal of approval. Now, Lysenko began dismantling Vavilov's Institute, piece by piece.

In the fall, Moscow hosted a nationwide agricultural exhibition. As director of the Lenin Academy, Lysenko was in charge of choosing the exhibitors and he had turned down several exhibits Vavilov thought ought to be there. Nikolai Ivanovich went to see Lysenko in his Moscow office.

Lysenko was sitting, with his eyes closed, at the end of a long narrow table. Vavilov started going down the list. Lysenko never said a word. If he approved a candidate he would slowly nod his head; if he disapproved he would do nothing. By the end, the list was cut down considerably. No one who had ever criticized Lysenko's position was accepted.

Vavilov could no longer contain himself. Several days later, at the agricultural fair, a worker "heard loud voices" coming from Lysenko's office. "Clearly sparks were flying. We entered the room and saw that Vavilov was holding Lysenko by his suit collar, cursing him for ruining Soviet science. Lysenko was terrified, he screamed that he, Lysenko, was an untouchable as a deputy of the Supreme Soviet, he would complain to the government and that Vavilov would be held responsible for an attempt to beat him up."[8]

Vavilov tried to keep working normally, but Lysenko confronted him at every turn. Vavilov was finishing his book *World Varietal Resources of Grain Crops*, but Lysenko closed down the Institute's publishing house, preventing its publication.

To colleagues, Vavilov still tried to play down the crisis. In a letter to a younger Institute researcher who was despondent about Lysenko's attacks, Vavilov wrote that there were "no threatening circumstances, so work quietly and organize your work [for publication] as soon as possible."[9] To his friend Hermann Muller, now safely outside Russia, he was more truthful. "The discussion is continuing," he wrote in a letter. "Our opponents are practically neo-Lamarckists . . . they have no experimental data . . . it is mostly a matter of faith. Nevertheless, it is considered Darwinian. The only way out for us now is to show more and more definitely the importance of modern genetics for breeding work."[10]

The recurring question for Vavilov, and, indeed, for all the geneticists, was the role that Stalin was playing in their persecution. On November 20, 1939, at 10 P.M., according to secondhand reports, Stalin sent for Vavilov, the only time he was given a one-on-one audience. When Nikolai Ivanovich entered the Kremlin office, Stalin was pacing back and forth, looking down, and he did not respond to Vavilov's entrance. Vavilov waited, then began to speak about the work of his Institute. Stalin continued to pace, saying nothing. After several minutes, Stalin sat down at his desk. "Well citizen Vavilov, how long are you going to go on fooling with flowers and other nonsense? When will you start raising crop yields?"

Nikolai Ivanovich began to explain his research and the importance of the world collection. Stalin listened briefly and then said,

"You may go." (There remains some doubt as to whether this interview ever took place because it is not recorded in Stalin's official visitor's book, recognized as a reliable record of Stalin's meetings.)[11] However, there is evidence that Vavilov believed Stalin was his main accuser, not Lysenko. A young geneticist, Nikolai Dubinin, recalled that one day in 1939 Vavilov had invited a group of geneticists to his Moscow apartment near Kursky Vokzal, the city's railroad station for the southeast, and the subject of Stalin's attitude came up.

"Do you know that Stalin is not pleased with me and that he supports Lysenko?" Vavilov asked, according to Dubinin. And Dubinin replied, "But Stalin keeps his silence, and this can be interpreted as an invitation to continue the discussion."

"Yes, maybe you are right," continued Vavilov, "but I am still under the impression that I and you and other geneticists often debate not with Lysenko, but with Stalin. To be in opposition to Stalin's views, even in the areas of biology, is not pleasant business."[12]

At the end of 1939, Lysenko sent Vavilov on a scientific expedition to the Caucasus. While he was away, Lysenko replaced the Leningrad Institute's entire scientific council—twenty-seven senior scientists. Vavilov was outraged. In a letter to the new agriculture commissar, Ivan Benediktov, he complained that Lysenko's move was "wholly unprecedented in the history of Soviet science," and represented "an intolerable settling of accounts."[13] Among those dismissed were his friends Georgy Karpechenko, director of the genetics lab, and Leonid Govorov, director of the department of grains. In the letter, Vavilov named the twenty-seven dismissed scientists, several of them party members. They had been replaced with people "with scientific views acceptable" to Lysenko.

"To circumvent the director on the highly important issue of membership of the professional council is strange, to say the least," he wrote, adding that implementation of Lysenko's decree meant that "the Institute cannot function normally." Vavilov again offered to resign because Lysenko's actions had "made it impossible for me to continue as leader of the Institute."[14] He "urgently requested" the commissar's intervention, but Benediktov did nothing. The commissar had already declared his hand: he stood with Lysenko.

As the weeks passed, Nikolai Ivanovich became increasingly despondent. In March 1940, he was in Moscow and asked one of his researchers to join him for an evening stroll. They walked slowly. After a few minutes Nikolai Ivanovich stopped and told the researcher that he was tired of life. He was ready to give up. He had "shaken up" a lot of science theory, and practice, he said, and it was probably time to stop.

The researcher would recall, "He took off his astrakhan hat, bent down a bit and spread his hands in a gesture of helplessness."

The researcher protested; it was no time to surrender. He had a long life to live and he could still do a lot.

It was dark and the researcher suggested they return to the Lenin Academy since it was already about ten o'clock.

"But he kept standing there, looking toward the east over the roofs of the Yaroslavsky railroad station. He said he was looking in the direction of Siberia and he talked about its natural resources and its massive, undeveloped farmland. Once again I tried to get him indoors, but he wanted to stay out. He complained about his health, that his heart was weak and his joints had started to ache. He was now fifty-two years old, and had rapidly become old for his age."

It was almost eleven o'clock before Vavilov agreed to go back to the academy where he said he would work for two or three hours and then go to sleep—on the sofa. He was used to it and, in any case, he would be up at five to work some more "in the quiet of the morning."[15]

In the last week in February 1940, Vavilov's sister Alexandra had fallen ill. At first it had seemed as though she had a bad cold, but her condition worsened and she became hospitalized. Nikolai Ivanovich visited her each day, as he had done with his younger sister, Lydia, when she had contracted smallpox and died in 1914. Alexandra's condition worsened and she died on April 3, two years after her mother. Sergei would write in his diary, "Such mystical dates . . . life feels so senseless in these minutes."[16] Of the original seven children only Nikolai and Sergei remained alive.

CHAPTER 26

The Arrest

Stalin was evidently looking for a moment when Vavilov's arrest would attract the least amount of attention, especially abroad. The opportunity came in the summer of 1940, the first summer of World War II, when the Soviet Union was still out of the war under the Nazi-Soviet Pact, and fighting between Germany and the Allies was consuming Europe. Nikolai Ivanovich left Moscow for his last plant hunting expedition, to the western Ukraine.

By the early summer of 1940, Vavilov seemed to accept the inevitable; he would be arrested. His life as a key voice in Soviet biology, perhaps even as an important scientist, was about to end. He tried to keep his anxiety from Yelena, but when he left for work in the morning, usually in a car driven by his chauffeur, he would call Yelena immediately when he arrived at the Institute, only a couple of minutes away. And when he had finished work, however late the hour, he would call her again to say he was on his way home.

At home in his Leningrad apartment, the conversation with

Yelena or visiting colleagues was dominated by his battle with Lysenko. His son Yuri, now twelve, had become a sensitive and studious youth who collected pictures of model cars and liked to read. Yuri would overhear his father in his study talking about Lysenko, often late into the night, with the staff from the Institute. Yelena was able to offer little relief. She was suffering from a nerve disorder that affected movement in her arms and hands, and she was officially registered as an invalid.[1] She performed household chores with great difficulty and, like Nikolai Ivanovich's first wife, Katya, she rarely cooked anything more than a bouillon cube.

At the Institute, the staff noticed Vavilov slowly weaken, his boundless energy gone. One of the researchers was working late one night. The main lights were dimmed and the darkened room was quiet. Suddenly the door opened and in walked Nikolai Ivanovich. "I had never seen him look like that," the researcher would recall. "Tired and silent, he sat down without removing his coat or hat, leaned his walking stick on the arm of the chair, and sat silent, for a long, long time. I rose quietly, heated the kettle, made tea and offered him a glass without a word. He drank the tea and went on sitting in silence. His face was contorted in agony. . . .

"Finally he rose, walked to the door, turned in the doorway and quoted Shakespeare's line, 'Ophelia, there is no truth on earth.' That was the evening, I later learned, after his last meeting with Molotov. That was when he finally saw everything clearly."[2]

But somehow Vavilov always managed to summon up new reserves of energy for his plants. It was early summer, the season for spending time at the experimental farm at Pushkin, a time to be in the fields at dawn, examining shoots of the new plants. The rosebushes were in bloom around the old Victorian house sent from England by Queen Victoria. Vavilov's strength, and his weakness, was that he could set aside everything—even his own welfare—in the name of science. As the summer wore on, Nikolai Ivanovich launched fearlessly into a new project of crossbreeding wheat and flax.

Despite Lysenko's power grab, Vavilov was still director of the Leningrad Institute and also the Institute of Genetics in Moscow. In his last letter sent abroad, on June 22, 1940, Vavilov wrote to S. C. Harland, the British cotton breeder, that "research at the Institute moves on," but "we are still caught in heated arguments with Lysenko's group—[who] follow the Lamarckists' point of view. We are on the opposite side—'conservative,' 'classic' geneticists."[3]

Even as his scientific life was about to end, Vavilov refused to let these battles divert his attention from the introduction of new varieties, his continuing taxonomic studies, and the publication of scientific monographs. His world collection now included 250,000 samples. In a 1940 paper entitled *Introduction of Plants During the Soviet Era*, Vavilov noted that between 1930 and 1940 his Institute had distributed up to five million packages of seeds to research, breeding, and other agricultural institutions. Two hundred and fifty-four new varieties had been put into production by the Institute and its research stations, half of them fruit and berry crops. Many of these varieties were specific to the climate zones of the Russian land-mass, from the Arctic to the subtropics, and included a wide range from staple grains such as wheat, rye, barley, and oats to medicinal plants. The quinine cultivation from the cinchona plants he brought back from South America had been a success.[4]

Since the early 1930s Vavilov and eighty of his collaborators had been compiling a comprehensive botanical review in twenty-two volumes entitled *Cultivated Flora of the USSR*. By 1940, seven volumes had been published, including unique monographs on grain cereals, legumes, small fruits and nuts, and fiber crops. The volumes were designed to serve as handbooks for plant breeders and bota-nists. In addition, he was preparing an international manual, *Theo-*

retical Bases of Plant Breeding, a joint work involving sixty authors. The first two volumes of 2,600 pages had been published.

Finally, he was preparing a third massive work establishing a new classification of cereals and legumes that would direct plant breeders to varieties available for hybridization. It had the grand title *World Resources of Cereals, Grain Leguminous Crops and Flax and Their Utilization in Plant Breeding*, but he had only written the first part.

Vavilov was still confident, apparently, that the Institute's presses would run again and he had not given up hope that "a collective work," a critical review of the main theoretical concepts of genetics, would be published. The titles of the planned papers—including "The True Status of Mendelism," "Michurin and Genetics," "Chromosome Theory from the Point of View of Dialectic Materialism"—showed that Vavilov had not surrendered. Of these works, he told Harland, "They are not meant for education purposes; they will be serious studies with critical analysis of former mistakes—and also give a plan for research required in the future." The collection was never published.

News from his foreign scientific contacts sometimes lifted his spirits. On May 4, 1940, Vavilov received a telegram from the U.S. Academy of Sciences inviting him to another world gathering, this time the Second International Congress of Pure and Applied Science of Physics, Chemistry and Biology at Columbia University in September. Vavilov sent a telegram to Molotov, the foreign minister, asking "for your instructions." Molotov did not reply.[5]

At other times, the news was grim. Another of his close friends, the Bulgarian geneticist Doncho Kostov, left for home. He had married a Russian woman, but returned to Sofia "for family reasons." Once there, Kostov arranged for Vavilov to be awarded an honorary doctorate of a Bulgarian university, and sent him an article that had been written about the award. Vavilov wrote back thanking Kostov "for the obituary."[6]

By the summer of 1940, Vavilov seemed to find renewed strength. Asked about the future of the Institute, he was, once again, optimistic. "If all our enemies were to be drowned in the Fontanka, they are so insignificant there wouldn't even be any bubbles," he told a member of the staff. (The Fontanka is a branch of the River Neva that flows by the Institute.) Asked whether he thought they would arrest him, he replied, "They won't dare."[7]

In his Leningrad office he had a plan of action, boldly entitled "Work Plan for 1940–41." It was a list of twelve books and articles he intended to write. Piled high on his desk were more than 2,500 pages of unpublished manuscripts with such titles as "Combating Plant Diseases by Breeding Resistant Varieties," which the Institute would submit for a Stalin Prize, plus some unfinished papers, "Field Crops of the USSR," "Plant Breeding in the Caucasus," and the biggest one of all, "Farming Regions of Five Continents," which described his expedition through fifty-two countries.[8] It was as though he wanted everyone to know that his work was not yet finished.

Meanwhile, the Kremlin plotted Vavilov's arrest. The NKVD produced an elaborate plan to keep the arrest quiet for as long as possible. If he were arrested at home, or even traveling on a train between Leningrad and Moscow, there would be witnesses. His disappearance would be known instantly, or certainly within twenty-four hours. The war had created an unexpected alternative.

The Commissariat of Agriculture wanted assessments of the farming potential of the western Ukraine, ceded to the Soviet Union under the Molotov-Ribbentrop Pact. In May 1940, Vavilov was appointed to lead an expedition to the new territory.[9] As he planned the route and selected the staff members, Vavilov seemed to sense that he might never return to Leningrad. He began to reassign those members of his staff that he thought might be in danger of losing their jobs if he were arrested. He judged, correctly, that Lysenko's first target would be the Institute of Genetics in Moscow, and he moved several researchers there to other departments. Of course, he could not explain his reasons and some did not understand and were upset, assuming he was displeased with them. In only a few weeks they would appreciate his protective instincts.[10]

On July 23, Vavilov's trip to Ukraine was finally authorized, and Beria drew up the papers for his arrest. Before leaving for Moscow Vavilov gave the staff a pep talk. Among them were his two favorite young researchers, Fatikh Bakhteyev and Vadim Leknovich. He was suddenly confident and excited again, making lists of equipment and spreading maps on the floor of his office. To those who had watched the elaborate planning of his foreign adventures in the 1920s, it was like old times.

"Ladies and gentlemen we have been entrusted with a most responsible task," he began his final talk to the staff.[11] For a brief moment, a month or so, he would leave behind the poisonous atmosphere of Moscow, and his beleaguered Institute in Leningrad, and do what he liked most, what he seemed born to do. Once again he would be collecting plants.

In Moscow, he was encouraged by a meeting he had with Andrei Zhdanov, a member of the Politburo for scientific matters. Zhdanov apparently left him with the impression that things would now take a turn for the better. After that meeting, one of the staff of the Moscow Institute of Genetics had complained to Vavilov that she had been removed from her project for "seditious" work—researching the action of X-rays on plants, the same line of experiments followed successfully by the American Hermann Muller. Vavilov reassured her that he had been talking to a person "high up in the government" and that she should not worry. She would be able to continue her work. "What I am telling you is something more than smelling salts," he promised.[12]

Before leaving at the end of July, Vavilov paid a last visit to

Lysenko. It did not go well. Their loud voices were overheard in
the hallway outside Lysenko's office. Vavilov stormed out, slam-
ming the door. A witness to the incident whispered to her colleague,
"Now, he's going to be arrested . . . he said such terrible things. He
told Lysenko, 'Thanks to you our country has been overtaken by
other countries.' You just wait, he'll be arrested."[13]

The expedition left Moscow by train, stopping three days in Kiev,
Lysenko's home territory. There, Vavilov held talks with local
agricultural bosses, visited a sugar beet research institute and an
archaeological exhibition, and spoke to a gathering of the Pioneers,
the communist youth group. The expedition then took off in three
black Soviet-made Fords for the journey south to Lvov and then
Chernovtsy. Along the way, they stopped to chat with local farmers,
examine their crops, take samples, and eat in local village cafés.[14]

Nikolai Ivanovich wrote Oleg a letter filled with excitement
and the promise of the new expedition. It was his last letter. "Dear
Boy! Today I am traveling to Bukovina, to Chernovtsy, from there
to the Carpathians. These places are beautiful. I crossed all Pod-
olia, Lvov and Ternopol districts. I will be on the road for another
couple of weeks and a half. There are some transportation difficulties, but I managed so far. I hope to master the philosophy of the
Carpathians."[15]

By August 5, Vavilov's expedition had reached the ancient
Ukrainian town of Chernovtsy where the group stayed the night in
a hostel attached to the university. Next morning, on August 6, he
was up at dawn, as usual. It was a splendid summer day, perfect for
plant hunting.

The expedition moved slowly up a dirt road into the Carpathian
Mountains, stopping to gather samples whenever Nikolai Ivanovich
spotted a rare grass. At dusk Nikolai Ivanovich and his band of bot-

anists returned to the hostel, pleased with the day's work and looking forward to discussing over the evening meal what rare grasses
they had found.

It was at this moment that the four NKVD agents stopped him
and told him that he was needed in Moscow. Whether he believed
them, or whether he realized immediately that his time had come,
we will never know. The NKVD records show that he was driven to
Lvov, from there to Kiev, and thence to Moscow. Somewhere between
Chernovtsy and Kiev, they stopped for a meal. The file records that
the agents gave Nikolai Ivanovich 10 rubles and 30 kopeks out of his
money they had confiscated to pay for his share of the meal—his last
meal outside prison. He arrived in Moscow on August 10 and was
incarcerated in the notorious inner circle of the NKVD prison system.
Like so many others of his class and promise, he had disappeared.

The day after Nikolai Ivanovich's arrest, NKVD agents searched his
apartment in Moscow, his apartment in Leningrad, and the dacha in
Pushkin at the Institute's experimental farm. Yelena Barulina and
Yuri were staying in Pushkin. As soon as the agents arrived, Yelena
sent Yuri to his room and told him to stay there. He remembers that
the agents came early in the morning and although he did not witness their search, he did see them drive off in black cars. Agents then
searched Vavilov's office at the Institute in Leningrad and also at the
Institute of Genetics in Moscow.

Among Vavilov's confiscated possessions from Leningrad,
according to the NKVD file, were 2,500 pages of scientific manuscripts in nineteen folders, a "brochure of which contents must not
be revealed," a "pistol with white metal coating," and "two combat
spiral-grooved bullets." From the apartment in Pushkin, they took
"various documents (telegrams and letters), notebooks with various
notes and addresses," and "one box of photo plates 9x12."[16]

During the raids the agents never mentioned his arrest and there was nothing about it in the newspapers. The Institute staff only learned of his arrest when his expedition colleagues Lekhnovich and Bakhteyev returned from Ukraine on August 12.

Sergei Vavilov immediately tried to find out what had happened. He wrote a letter to the prosecutor's office jointly with the Academy of Sciences president Vladimir Komarov, but there is no record of the letter. Komarov was appalled that Vavilov had been arrested "because he had the courage not to agree with Lysenko," but he was not a fan. He was jealous of Vavilov's international reputation, and he may have balked at sending the letter for fear of Stalin's reaction.[17]

On August 13, Sergei Vavilov wrote in his diary, "He is in Lvov now. This means that the arrest will thunder, this means that his big useful life is being ruined, his and the lives of his close ones. For what? The whole life of tireless and intense work for his homeland, for the people. All his life spent in work, with no other hobbies. Wasn't it obvious and clear to everybody? What else can be asked and demanded of individuals? This is a cruel mistake and an injustice. It is even more cruel because it is worse than death. The end of scientific work, the slander, ruining the lives of family members, the threat of it all.

"My diary is full of grief: mother's death, sister's death, and now the horror hanging over my brother. I cannot think of anything else. It is so terrifying, so pitiful, and it makes everything senseless. Good thing that mother died before this, and it's too bad I failed to die too. It's unbearable torture."[18]

Vavilov's old professor from the Petrovka, Dmitry Pryanish-nikov, then seventy-five, courageously called the Prosecutor's Office on August 14, and confirmed that Vavilov had been arrested. But he was given no details.

In Moscow, Vavilov's son Oleg, now twenty-one years old, had received the last letter that Nikolai Ivanovich had written. He could not believe that his father had been arrested. It was a "mistake," he kept repeating to his mother, Katya. "Father could not have committed any crime. Can Lysenko be such a scum?"

Oleg would sit at the dining room table holding his head in his hands, muttering, "I pity Dad so much," and repeating over and over, "He couldn't have done anything wrong."[19]

The atmosphere in the Moscow apartment was tense, with Katya repeatedly declaring that she had told Nikolai Ivanovich that he would be arrested if he failed to "subdue his pride." He should have followed the example of his younger brother, Sergei, not have caused a stir and not have confronted Lysenko, and then nothing would have gone wrong. In reply, Oleg would only throw his hands in the air.[20]

Oleg asked the NKVD where they could send food and clothing and they were finally told the NKVD's jail, the dreaded Lubyanka. They were told to bring money, 30 rubles, and underwear. They took packages regularly, handing them in at the grim reception room, but they would never know if he received them.

The arrest had left Oleg, then a student at Moscow University, without money—and Katya was unable to provide for him. He wrote to the interior commissar, explaining that he was in "dire financial straits." He asked for permission to get his father's salary for the month of August. To do so he boldly requested his father's "power of attorney." There was no reply.

In September, Sergei Vavilov fell into a deep depression about Nikolai, about the war, and about all the talk of science and practice. To him, all science had practical goals and the debate about theory was "pointless." "I would like to die suddenly without feeling it," he wrote in his diary.[21] He would look into the glass of his desktop and see Nikolai in his own reflection. "Like a ghost. It's so scary," he wrote. He felt he had suddenly aged. "Until now I thought of myself as almost a teenager. Now I am aging. I feel total loss of creative stimuli, helpless, talent-less and weak. People are barely distinguishable from grasshoppers, or cars. . . . I am turning into a lifeless object and to live in these conditions is a hard task."

For his brother, he knew, life was even worse.

CHAPTER 27

The Interrogation

Over the next eleven months Vavilov would be interrogated almost four hundred times for a total of 1,700 hours. Some of the sessions lasted up to thirteen hours. His chief interrogator, a thirty-three-year-old lieutenant named Alexander Khvat, had been an official with the Komsomol, the party's youth organization. Khvat had worked in the jails for two years and was known for his extreme methods. He began his interrogation on the day Vavilov arrived at the jail on August 10 at 11:35 P.M. and finished at 2:30 the following morning.[1]

Stalin had finally put Nikolai Ivanovich where Lysenko wanted him, in jail.[2] Vavilov knew perfectly well why he had been arrested. The list was long—for his opposition to Lysenko's false doctrines and anti-science speculations, for his skepticism and halfhearted support of Stalin's agricultural policies, for his attempts to free his repressed colleagues, for his contacts with foreign scientists, and for his fame within the international scientific community. But these were not criminal offenses. The NKVD had to fabricate a case against him.

On August 12, two days after his arrival in the jail, his interrogator, Lieutenant Khvat, laid out those false accusations for the first time. They were brought under Article 58, "State Crimes," of the Soviet Criminal Code, which included "Treason to the Motherland—acts that might damage the military might of the USSR, its national sovereignty or the inviolability of its territory"; "Wreckage"—undermining the economy; and "Sabotage," including "inhibiting activity in the interests of capitalist organizations, destruction for counter-revolutionary purposes by use of explosives, arson of railway warehouses."[3] For these, Vavilov faced death by firing squad.

For the next eleven months, the NKVD produced a typewritten record of the almost four hundred interrogations. These records consist of question and answer. The questions are often insulting and chilling, evidence of Khvat's brutality and the suffering of Vavilov. In some parts, the transcript is difficult to read because of Lieutenant Khvat's barbarism. The record is said to be the most complete of any released by Soviet state security for this period.[4]

There is no way of knowing how faithful the official transcript is to the original, whether the answers are direct quotations, summaries, or fabrications. Some of the answers have an authentic ring about them and contain opinions that Vavilov certainly held and phrases that he might have used. But the NKVD had specialists who wrote up false transcripts.[5] Ten years after Nikolai Vavilov's "rehabilitation," the file was first seen by the Russian journalist Mark Popovsky in 1965, when he was allowed only to take notes. The entire investigation file, No. 1500, was given to Nikolai's son, Yuri Vavilov, and selected texts were published in Russian for the first time in 2004.[6] According to this NKVD record, Khvat began his interrogation with these words:

KHVAT: You are arrested as an active participant of an anti-Soviet wreckage organization and a spy for foreign intelligence services. Do you admit your guilt?

VAVILOV: No I do not admit my guilt. I never was a spy or participant of an anti-Soviet organization. I always worked honestly for the benefit of the Soviet state.

K: You are lying. The investigation is aware that during a long period of time you headed the anti-Soviet wreckage organization in the field of agriculture, and you were a spy for foreign intelligence services. We demand truthful information.

V: I categorically declare that I was not involved in espionage or any other anti-Soviet activity.

K: The investigation knows you as a man who is principally hostile to the existing regime and the policy implemented by Soviet power, particularly in the area of agriculture.

V: I maintain that the investigation possesses one-sided materials that cast an erroneous light on my activity, and evidently they result from my disagreements regarding scientific and administrative work with a whole number of people who have described my work tendentiously, in my opinion. I maintain that it is nothing but slander.

K: The talk here is not about your differences of views with some scientific employees, but about your active anti-Soviet work. . . . We suggest you give serious thought to the questions posed by the Investigation, and give information on the essence of the charges presented to you.

V: I was not involved in anti-Soviet activity.

On August 13, Vavilov was fingerprinted and photographed. A week after his arrest, this robust middle-aged man is unshaven, exhausted, and the sparkle has gone from his eyes.

For the next two weeks he was interrogated mostly at night. Vavilov never wavered from his original denials. And he even wrote a letter to Khvat's NKVD boss, Lavrenty Beria, declaring that he had never betrayed his Motherland, had never engaged in any counterrevolutionary activity, had always devoted himself to scientific work, and had never spied for a foreign power.

Then, suddenly, after an interrogation that lasted for nine hours and fifty minutes during the night of August 24–25, Vavilov admitted, "I am guilty of being a member of a rightist organization existing in the system of the USSR Commissariat of Agriculture since 1930."[7]

He still maintained he was innocent of espionage, but Khvat warned him, "You will not be able to conceal your active espionage work, and the Investigation is going to interrogate you about it."

Why Vavilov made this "confession" is a matter of speculation. Was he tortured and could not stand the pressure? Or did he decide that being a member of this unnamed "rightist group," a fictitious organization, was a lesser charge for which he would not be executed but rather sent to a labor camp? It is also possible he considered that his admission on this charge could relieve the pressure on his investigator to extract a full confession on all charges.

What is known, however, is that from the end of August, the record of the investigations began to appear in printed form, rather than notes, suggesting that they were now receiving wider circulation, presumably to high echelons of the party.

The NKVD interrogators were always under pressure to bring out a full confession, and Khvat was under a special burden. When he was appointed to Vavilov's case he asked to be excused, saying he knew nothing about agriculture, but his request was denied. This meant his career was on the line if he failed.

There were no restrictions on Khvat's methods. The systematic torture of political prisoners that began during the Terror of 1937 continued and was specifically authorized by Stalin. "It is known," Stalin had written at the time, "that all bourgeois intelligence services apply physical pressure against the socialist proletariat and apply it in a vast variety of forms. Hence the question: why would the socialist intelligence service be more humane towards inveterate agents of the bourgeoisie, inveterate enemies of the working class and kolkhoz-members?"[8]

Importantly, two other NKVD agents were involved in Vavilov's interrogations. One of them, a Major Shwartzman, was the head of the Chief Economic Department and would later be revealed as a persistent torturer and writer of false transcripts.[9] His "blood trail" would be found in the pages of many interrogation files that were eventually released half a century later.[10] Another was a poorly educated agent named Albogachiev. The heaviest period of interrogation during the eleven months—twenty-three interrogations totaling 120 hours—was between his arrival in prison on August 10 up to the night of August 24 when Vavilov "confessed." For ten of those nights he had to stand in front of the interrogator and try to keep awake.

There is some evidence that Nikolai Ivanovich worked out a strategy for dealing with Khvat that gave more time for those outside to help him. His connections in high offices in the Soviet Union, such as his old professor Dmitry Pryanishnikov, and the international community, he must have hoped, would be working for his release, or at least a reduced sentence. Moreover, his early "admissions" were not confessions of guilt to criminal activity— "wrecking" and "sabotage." Rather, they were refusals to carry out damaging directives issued by his superiors—like Yakovlev, the commissar of agriculture, who had already been executed. In addition, Vavilov's "admissions" were more self-critical lapses of judgment and political correctness rather than deliberate sabotage. He spoke of not paying sufficient attention to details, of being hampered by his bourgeois outlook on life, of not paying enough attention to the Marxist "unity of theory and practice," and of not being vigilant about the choice of staff for his Institute.

When Khvat asked him to describe the "connections" in his "anti-Soviet activity," Vavilov listed a dozen leading figures in Soviet agriculture, such as Yakovlev, all of whom had been shot or were under arrest. It was Yakovlev, he said, who had recruited him into the anti-Soviet group.

Khvat immediately asked, "Why . . . what were his [Yakovlev's] reasons?"[11]

V: [Yakovlev] came to learn of my anti-Soviet attitudes, which at the beginning were most clearly expressed in my high evaluation of the American and West-European agriculture, stressing its advantages compared to the development of agriculture in the Soviet Union. Besides, to a certain degree, I favored developing strong individual peasant's farming.

K: That is, of kulaks?

V: Yes.

When Khvat asked about the "direction" of his "hostile activity," Vavilov said that it "consisted mostly" of the "gap between scientific and practical work," of ignoring "the development of experimental work," of incorrect distribution of a number of varieties, including corn and cotton, and disruption of work to organize crop rotation.

To answer these questions more fully Vavilov asked for more time—"for a chance to recollect all the facts of hostile work carried out by me and co-participants known to me, and to present them to the Investigation in detail during future interrogations."

Khvat responded sharply, but positively, warning, "Don't try to explain it with blunting of your political vigilance . . . we demand truthful testimony."

. . .

Having failed to extract a confession of espionage, Khvat seemed determined now to obtain as much information as possible about Vavilov's "co-conspirators." In September, Khvat apparently received specific instructions to broaden the scope of his interrogations. He wanted names, dates, and details. It is possible that Khvat had been ordered to prepare for a show trial.

For a while, Vavilov avoided direct answers, again emphasizing his own bad policy decisions and the laziness of ivory tower academics rather than deliberate sabotage. As director of the Institute, he took full responsibility for what had occurred—without actually admitting blame. And, as a counterbalance to his failings as a director, he pointed to the success and the "huge richness" of his seed collection "with no equal in the world. . . . It contains all the best standard and local varieties both for our country and other countries with similar climates."

Some of these seeds, he admitted, had been ruined, because they were not propagated in time due to the "wrecking activity" of certain staff researchers. He could not "fight this wreckage directly" but, he admitted, he should have removed those who were guilty.

Khvat soon tired of these evasive answers and increased the pressure. At the beginning of October, he was joined in the interrogations by his assistant, Albogachiev, and together they extracted a confession from Vavilov that he had been a member of the counter-revolutionary Peasant Labor Party. As this was known to be, even then, a purely fictitious group, it is still possible that Vavilov did not think that this confession amounted to much. It is also possible that he confessed after hours of unbearable torture.

Evidently, Khvat did not consider the confession enough for his purposes. He accused Vavilov of "concealing many facts" and demanded he now "testify truthfully." He started to ask Vavilov for names.

"Who did you protect from the investigation?"

Khvat already had a list of names in Vavilov's operational file, but, according to the interrogation records, Vavilov named six researchers including Institute professors and personal friends. Among them were Leonid Govorov, a colleague from student days and head of legumes at the Institute, Georgy Karpechenko, then aged forty-two, a plant geneticist and expert on chromosomes, and the older Konstantin Flyaksberg, aged sixty-one, the head of the wheat section, who had worked in the old Bureau of Applied Botany. Govorov had been one of the first of the Institute members to go to Moscow to plead unsuccessfully for Vavilov's release. Karpechenko had written a report after Vavilov's arrest saying work at the Institute could not confirm Lysenko's theories. Govorov and Karpechenko were arrested in February. Flyaksberg would be arrested six months later.

According to the record, Vavilov said, "I chose these people to work at the Institute taking into account their closeness to my anti-Soviet views and to the adversary direction of my work. These people always fully supported me in the Institute's wrecking activity that consisted in the deliberate gap from practical tasks."

The interrogation methods would always dangle hope at the end of some confession, but there is no way of knowing if Khvat had offered an enticement to Vavilov, a lesser sentence, perhaps.

Certainly after months of these sessions, Vavilov must have begun to despair that the interrogations would still end with his execution. And as the year turned, he must have imagined that he would never see his family again.

For his brother, Sergei, 1940 had been the "hardest" of his life. "Hard in futility, in stupid cruelty," he wrote angrily in his diary. "I'm beginning to look at the future in the same simple, calm and cold-blooded way as a stone on a dusty road . . . petrification, fossilization, is a means of self-defense."[12]

•   •   •

By the spring of 1941, the record shows that Khvat had extracted another admission. Vavilov said he had built "an anti-Soviet ring" in 1931–32 at the Institute and that it included, among others, Govorov, Karpechenko, and Flyaksberg. Now these three close colleagues became members of a "wreckage ring."[13]

To exploit this new admission, Khvat now employed one of the most sadistic techniques in the interrogation methods book. He staged face-to-face meetings between Vavilov and the three men. All the participants knew was that one of them had betrayed the other. Khvat brought these men, one by one, into the interrogation room and made them face Vavilov, once their friend, their boss, and now, perhaps, their accuser. One can only guess at the emotional torment of these sessions for Vavilov, who was already mentally and physically exhausted. His friends had also been through the NKVD grinder and this part of the record is especially distressing.

Perhaps the most upsetting is the confrontation with Georgy Karpechenko because it was Nikolai Ivanovich who had persuaded him to return to Russia from America. In this confrontation, Khvat even constructed a situation where the two men argued with each other about what the other had said.

QUESTION to Vavilov: You testified that in conversations with you, both at work and in your apartment, Karpechenko expressed anti-Soviet sentiments and was praising conditions of capitalist countries. Is this correct?

VAVILOV: Correct.

QUESTION to Karpechenko: Do you confirm this part of Vavilov's testimony?

KARPECHENKO: Yes I confirm.

QUESTION to Karpechenko: You testified that you were recruited by Vavilov to participate in anti-Soviet activity only in 1938, and

Vavilov testifies, and you also confirmed it during the face-to-face interrogation, that the anti-Soviet activity connection was established between the two of you in 1931–2. Why do you conceal the true date?

KARPECHENKO: I insist on my testimony that Vavilov recruited me for anti-Soviet work in 1938.

Vavilov was then asked to "specify" the date and he replied: "I consider that 1931–32 will be more exact."

QUESTION to Vavilov: Do you have questions for Karpechenko?

VAVILOV: I have no questions.

QUESTION to Karpechenko: And do you have questions to Vavilov?

KARPECHENKO: I don't have questions to Vavilov.

We do not know if these sessions were accompanied by physical torture, but the mental strain was torture enough. This was how a close working relationship between the two brilliant scientists was broken; how a firm friendship ended between a teacher and his pupil, a professor and his student, an administrator and a head of department, two people who had shared a common scientific dream, who had believed in the socialist utopia—at least to the extent that it would provide them an unrivaled opportunity to achieve their scientific goals for humankind.[14]

As the investigation into Vavilov continued, it became clear that Khvat's latest orders were to wring out of his victim as much information as he could about as many specialists as possible in Vavilov's plant breeding empire—information that was of particular interest, of course, to Lysenko as he annexed the fiefdom that once belonged to Vavilov. Each time a geneticist was arrested it was one fewer in Lysenko's way, but there is no document in the official dossier showing Lysenko had intervened in the interrogation—at this stage.

In the end, Khvat could not break Vavilov on the key charge of espionage. Vavilov's resistance would become a source of pride among his admirers. Vavilov, these admirers noted, did better than Galileo, who was accused of heresy by the Vatican for believing that the sun, rather than the earth, was the center of the universe, and under threat of being burned as a heretic, recanted his ideas. Vavilov did not renounce his belief in genetics and in Mendel.[15]

There are many ways of breaking down a political prisoner's defenses, and the NKVD interrogators were unyielding. During eight months of interrogation, Vavilov shared a cell with a prisoner named Lobov. He was an NKVD informer, a plant, a prisoner who was given the opportunity of freedom for "singing" on his cellmates. Lobov was well suited to the task. He was a former assistant to the chief of state security in Leningrad, but he had been in jail since 1935, convicted for his alleged role in the murder of Sergei Kirov, the Leningrad party secretary, who was shot dead by an assassin on December 4, 1934. Vavilov, who had been supported by Kirov, would obviously have an interest in this man's version of Kirov's murder and possibly even have been persuaded that Lobov was on the side of the geneticists. Kirov's murder had never been solved and it is almost certain that he was killed on Stalin's orders. Lobov's report on Vavilov was filed in March 1941, and he was freed at the end of April.

According to the NKVD file, Lobov reported that Vavilov was "a man of exceptionally anti-Soviet attitude," a man who was "showing special hostility to party leaders and to the government."[16] Lobov allegedly said, "Vavilov continues to be a determined bourgeois scientist who even here in jail has not sheathed his sword that he holds against Soviet power. . . . He pretends to be an innocent victim in front of the investigators, but after returning to the cell he continues to speak against the leaders of the Party and the Soviet people."

Lobov's testimony is, of course, as suspect as the rest of these documents, perhaps even more so. He was buying his freedom and obviously telling his masters what they wanted to hear. But some of the phrases Lobov uses in his report could indeed have been spoken by Vavilov. He may well have called Lysenko a "false scientist." And he did think highly of Bukharin as a politician. And he might have called Michurin an "experienced" man of practice who had nothing to do with true science; that was how Vavilov viewed the old gardener of Kozlov. He also could have said that Stalin and Molotov were "ordinary human beings, the same as others, not the Gods that reptilian hallelujah-singers say they are."

In the early summer of 1941, Khvat was ready to close his interrogation, but before he did his superiors wanted something extra to bolster the prosecutor's case, a kind of insurance against being caught not understanding the science. Khvat set up an "experts commission" to review the scientific implications of Vavilov's responses and to make sure the record made sense to scientific minds. Khvat's immediate boss, Major Stepan Shundenko, was in charge. He was the same Shundenko, formerly a researcher at Vavilov's Institute, who had been a secret police informer.

This "experts commission" was a sham, and provided evidence of Lysenko's hand in the case. Shundenko drew up a list of five referees, all of them personally "approved" by Lysenko. Of the five men, four would later be described as "hostile" to Vavilov and one of them "simply hated him." The report, of thirty-one pages, fully supported the charges of wrecking against Vavilov. Instead of using its plant resources to improve agricultural productivity, Vavilov's Institute used them to justify "metaphysical and anti-Darwinian concepts," and the Institute was "always remarkable" for its "high concentration of socially hostile elements." By 1940 it still had "twenty-one employees of noblesse origin, eight from the priesthood, twelve from honorary citizens [tsarist professional class], ten from the merchant class."[17]

Fourteen years later, during Vavilov's "rehabilitation," one of these "experts" would admit to the fraud. He was Ivan Yakushkin, the botanist who had been working under cover for state security as an informer on Vavilov since 1930 and was among the first to provide information for Vavilov's police dossier. At the time of his appointment to the "experts commission," Yakushkin held a high position at the Lenin Academy, under Lysenko. He would describe how the "experts" never wrote a report; a text of the commission's report was prepared by someone else—presumably Major Shundenko— and read out to the "experts," who promptly signed it. Yakushkin would admit to the sham; he signed the report even though he did not agree with parts of it. "This was the kind of situation when it was frightening to refuse to sign," he would later admit.[18]

But the war intervened in this trickery. On June 22, 1941, Germany invaded the Soviet Union. NKVD investigators were ordered to close down cases under investigation. Khvat rushed through the last of the face-to-face interrogations. In those especially harrowing days, there is one account of Vavilov's physical deterioration. It comes from an artist, Grigory Fillipovsky, who would claim he was in the same prison as Vavilov, and even in the same cell. He recalled, "Every night Vavilov was taken off for questioning. At dawn a warder would drag him back and throw him down at the cell door. Vavilov was no longer able to stand and had to crawl on all fours to his place on the bunk. Once there his neighbors would somehow remove his boots from his swollen feet and he would lie still on his back for several hours."[19]

On June 29, Khvat's dossier was complete and ready for the

prosecutor. The only remaining task was to select exhibits from the material confiscated during the raids on Vavilov's apartments in Moscow and Leningrad, the office at the Institute, and the apartment at Pushkin. The items chosen included a 1918 manifesto of the Great Russian Union, a slavophile group to which Vavilov's father belonged; a photograph of Alexander Kerensky, the head of the provisional government before the October revolution; a 1922 brochure, *Breads in Russia*, by Robert Regel, the director of the old Bureau of Applied Botany, with an introduction by Vavilov where he had expressed fears for the future of Russian science if the bourgeois specialists were persecuted; a notice from the Lenin Academy president warning Vavilov about discussing Soviet agricultural issues with foreign scientists; and copies of two petitions on behalf of arrested members of the Institute.

All the other items, including priceless material from Vavilov's foreign trips—in ninety-two folders—ninety field notebooks, photographs, books, magazines, journals, maps, and letters, had been destroyed.

On July 9, three generals of the Military Collegium of the Supreme Court of the USSR took less than five minutes to find Nikolai Ivanovich Vavilov guilty of all counts against him, including espionage. The generals had satisfied themselves that "from 1925 Vavilov was one of the leaders of an anti-Soviet organization of right-wingers operating within the Commissariat for Agriculture and several scientific institutions in the USSR. . . . In the interests of these anti-Soviet organizations, he carried on widespread wrecking activity aimed at disrupting and destroying the collective farm system and the collapse and decline of socialist agriculture in the USSR. . . . Pursuing anti-Soviet aims he maintained contacts with White émigré circles abroad and transmitted to them information containing state secrets of the Soviet Union."[20] The generals called no witness and there were no lawyers present, no voice on Vavilov's behalf except his own.

Vavilov would later declare categorically that the charges were "based on hearsay, untruths and slanders which had not been confirmed to any degree by the investigation."[21] But he was sentenced to "suffer the supreme penalty—to be shot—and for all his personal property to be confiscated." The sentence was "final and not open to appeal."

In fact, there was one option still open to any Soviet citizen. Vavilov could appeal to the Presidium of the Supreme Soviet, the only government body with the power of pardon. That evening, he was issued a sheet of paper, a pen and ink. In a firm hand, he wrote that he "humbly entreated" the Presidium to "grant me a pardon, and to give me an opportunity to atone by my work for my guilt in the eyes of the Soviet regime, and the Soviet people.

"I pray that I may be granted the most minimal possibility of completing work for the benefit of the socialist agriculture of my Motherland. As an experienced educator, I pledge to give myself entirely over to the training of Soviet specialists. I am 53."[22]

On July 26 his appeal, number 283 in a list of 361, was turned down and he was transferred to Butyrskaya prison for the death sentence to be carried out. Vavilov's colleagues Karpechenko and Govorov were tried at the same time and immediately executed.

From death row, Vavilov wrote a personal plea to Beria for mercy. A few weeks later, Beria's envoy visited Vavilov in jail to let him know the Presidium of the Supreme Soviet had agreed to reduce his sentence. He was transferred back to the inner jail of the NKVD where Beria's envoy visited him again, on October 5 and 15, to discuss how his expertise could be used. He was told that arrangements would be made within a few days. But three hours after the envoy left, the war intervened again. All prisoners in his jail were rounded up, taken to the train station, and evacuated to eastern cities. The Germans were at the gates of Moscow.

CHAPTER 28

Return to Saratov

Nikolai Ivanovich's extraordinary life would end in 1943, not in the cities where he made his reputation as one of the great scientists of the twentieth century, not in Moscow, or Leningrad, or any of the distant places on five continents where he had risked his life for samples of rare food plants—in the Hindu Kush of Afghanistan, the high plateau of Abyssinia, the Andes of Argentina, Bolivia, and Peru. He would die of starvation in Saratov, in a fetid prison, on the edge of the rich black earth of the steppe. It was here that he planned to use his new plant varieties to end the ghastly cycle of Russian famines, and turn them into crops that would feed the world.

Nikolai Ivanovich first came to Saratov in 1917 as a young and proud professor of agronomy. While the civil war had raged around him, he had calmly explored new horizons of botany and geography, and, at the same time, he had found a new and enduring love in his beautiful, delicate Lenochka. He was proud of his science, and of his country, overly proud for his own good, as Katya had lectured him. Blinded by his devotion to science, he had refused to see the evil in those who sought to destroy him until it was too late.

Now, Vavilov returned to Saratov at the end of October 1941 as a convict, a political prisoner, battered and exhausted from Khvat's interrogations. Like his fellow convicts, he was emaciated, living in filth, riddled with dysentery and other disorders that would take many of their lives. His enemies had destroyed him physically, but they could not crush his spirit. Nikolai Ivanovich would live for more than two years in a prison where others like him, convicted of fabricated crimes, existed barely from day to day, and died suddenly. By chance, his beloved Lenochka and their son, Yuri, were also evacuated from Leningrad to Saratov. They would spend the rest of the war years searching in vain for a husband and a father who was in jail, dying, not fifteen minutes' walk away.

The evacuation of thousands of political prisoners from Moscow on the night of October 16, 1941, is a story now well known for its astonishing cruelty. Prisoners were brought to the Kursk railroad station around midnight, lined up in rows outside, forced to their knees, and ordered not to look up. In this position, they would spend the next six hours waiting for trains that would take them east, away from the advancing Germans, to Saratov, Orenburg, and Kuibyshev.

The first snow of winter had fallen, quickly melted, and left icy puddles. Those prisoners who tried to crawl out of the puddles were kicked back into line. When dawn came, the panicked citizens of Moscow, preparing to evacuate before the advancing Germans, began to appear on the street. In front of their eyes they saw for the first time the state's political prisoners who had been shut away, out of sight, now bowed and huddled together, cold, wet, and hardly alive. These were the counterrevolutionaries the citizens had heard about, the "wreckers" and the "saboteurs" of the Soviet state, the "enemies of the people" now exposed for all to see. The citizens of

Moscow did their patriotic duty. They hissed and yelled at the rows of prisoners, "Spies! Traitors!"

When the trains were finally ready for loading the prisoners, the guards forced twenty into compartments made for five. The journeys would last up to two weeks, and many would die on the way. The train carrying Nikolai Ivanovich was originally destined for Orenburg, but was forced by German air raids to divert to Saratov.[1]

Vavilov was taken to Prison No. 1, a red-brick tsarist building on Astrakhan Street. The locals called it "Titanic," because it was long and thin and had cells on six floors with tiny windows, like portholes. It was terribly overcrowded. Vavilov was put in a cell for the condemned in the windowless basement. His cell mates were a well-known philosopher, Ivan Luppol, and an engineer from Saratov named Ivan Filatov. Filatov's "crime" was to have had an uncle, a man of the bourgeois class who ran a timber wharf in the city before the revolution. To be of the bourgeois class was to face a death sentence in the Soviet state. Filatov's term was eventually commuted to ten years, but he was so ill that he would be sent home to die. The enterprising Russian journalist Mark Popovsky found a Saratov truck driver who had heard Filatov describe the conditions on death row.

The cells were narrow with a single bed and a table, each fixed to the wall. A single electric light bulb burned in the cell day and night. They would take turns sleeping two at a time on the bed, and one dozing at the table. They were dressed in canvas sacks with holes cut for the head and arms and they wore slippers made from the bark of a lime tree. They received food three times a day: two spoons of kasha in the morning, a tin cup of soup made from rotten salted tomatoes with a piece of bad salt fish for the midday meal, and a spoonful of kasha for supper. They were supposed to have three hundred grams of black bread made from barley flour, but there was often a fight over the bread, the strongest taking all. The condemned prisoners were given no exercise, and not allowed to receive parcels.[2]

Despite the conditions, Nikolai Ivanovich would still muster enough inner strength to organize life in the cell so that each day had a purpose, a plan to prevent himself and his two cell mates from giving up and dying. Although they were only supposed to talk in whispers, Vavilov arranged lectures, given by each of them in turn, on history, biology, and the timber industry. Vavilov would also tell stories of his adventures in foreign lands.

For a while, this made-up routine kept Vavilov and his companions from losing heart, and allowed Vavilov himself to continue hoping that his case was being reviewed. He had to think that eventually Beria would arrange for him to be transferred to a work camp, as Beria's envoy had said would happen before he was evacuated from Moscow. But when he explained to the governor of the prison that he had already received this assurance, the governor said he had no powers to move Vavilov out of the condemned cell. "If we get a paper from Moscow telling us to shoot you, we shall shoot you; if they tell us to pardon you, we shall pardon you," the governor had said.[3]

Fifteen minutes from the Saratov jail, Yelena and Yuri were living in one room of a one-story brick house owned by her sister, Polina, and Polina's husband, Maxim. He was head of the university's extramural education department and they had a son who was a geology student there. Yelena and Yuri had arrived at the beginning of September when Nikolai Ivanovoich was still in jail in Moscow. The journey from Leningrad had been fraught with problems. They were now officially classless citizens; the family members of an enemy of the people were also enemies of the people. They had no money, and no means of making any since enemies of the people were not employable. In addition, Yelena's disability pension for her nerve disorder was canceled.

When Vavilov was arrested, Yelena and Yuri had been living in Pushkin, in the flat near the experimental farm, but in the summer of 1941, after the German invasion, the farm and the Institute in Leningrad were evacuated. They had watched as Nikolai Ivanovich's priceless seeds were put into crates in preparation for shipment to the east by rail. Staff members approved by Lysenko were to be sent with the material, but that did not include Yelena and Yuri, of course. They could have stayed on, hoping the Germans would not reach the village, but Yelena was worried that they would be forced to leave with the other evacuees and sent into exile, who knew where?

Galina Karpechenko, the widow of Georgy, offered them accommodation in her brother's dacha outside Moscow.[4] Galina, sixteen years younger than Yelena, was in good health, and she was able to help Yelena do the laundry and cook.

Yelena made several trips to Moscow, taking food parcels to the Lubyanka and trying to find out more about Nikolai Ivanovich. As with Katya and Oleg's packages, the guards took the parcels and said nothing. From the dacha Yelena and Yuri watched the first German air raids on Moscow. The night sky was lit up and Yuri climbed a tree at the dacha to get a better view. No one knew how rapidly the Germans were advancing, but as a precaution they dug a small shelter in the Karpechenko dacha garden. Toward the end of the summer, Yelena's brother, Konstantin, who lived in Moscow, was able to arrange train tickets for them to leave for Saratov.

Polina's house had a vegetable plot and Yelena's cousins lived in the house next door. Yuri went to Saratov school for boys No. 21 and they all prepared themselves to live out the war.

In jail, Nikolai Ivanovich managed to find reserves of endurance that kept him alive, but there must have been moments of terrible despair about his family and his life's mission, now suddenly curtailed. He also worried about his Institute and what fate might have befallen his colleagues.[5] He did not know that his close friends Govorov and Karpechenko had been shot and he must constantly have wondered what had happened to them.

As he had feared, Lysenko had moved swiftly, and brutally, to take over the Leningrad Institute. He had fired specialists who were not his followers and put in a new director. Any geneticist who remained was subjected to severe criticism in open sessions and rudely shouted down. A few people suffered nervous breakdowns and an elderly professor became so agitated when they tried to take his office away that he declared he could not go on living. He went home and an hour later he died.[6]

By the end of 1941, Vavilov's health had deteriorated significantly, and there was still no word of a pardon from Beria. Vavilov had developed a bad case of scurvy and was sent to the prison infirmary. On the way he met another inmate, a sixteen-year-old schoolgirl named Irina, who was serving time for "trying to organize an attempt on the life of Comrade Stalin."

That day, the prisoners were waiting in the courtyard for transportation, standing facing the wall with their hands behind their backs. Irina was frightened and sobbing because she did not know why she had been selected to join their sickly group. The old man next to her, she would recall later, was wearing a black overcoat and was very thin. In a calm voice, he asked her, "Why are you crying?" When she told him, he said, "Listen to me carefully. Since you will almost certainly survive this, try to remember my name. I am Vavilov, Nikolai Ivanovich, an academician. Now don't cry and don't be afraid, we are being taken to the hospital. They have decided to treat me before they shoot me . . . don't forget my name." Then they were taken to the hospital and Irina never saw Vavilov again.[7]

A few weeks later Vavilov was returned to his cell and in the spring was allowed to write another plea to Beria. On April 25, 1942, he reminded Beria of the talks with the envoy about a pardon and the transfer to a labor camp. Vavilov stressed that he would be "glad at this difficult time to be used to improve the country's defenses as a plant breeder."[8]

While in prison in Moscow, he told Beria, he had written a book, "The History of Worldwide Agriculture," with the main focus on the USSR. (Apparently he had been given paper and pencil and had produced many pages, none of which survived.) And before his arrest he had finished "a large work" on plant diseases and several other books on plant breeding, including the book about his expeditions on five continents.

He said he was "quite strong physically and morally" although the inmates were suffering at the time from an epidemic of dysentery and many were dying. Otherwise he did not complain about the miserable conditions in the prison. "I request and beg you to make my lot easier, to decide on my future and to allow me to work in my special field, even at the lowest level," he concluded.

When Vavilov's letter eventually arrived on Beria's desk it was not the only communication about his case. On April 23, two days before Vavilov had written his plea, he was elected a foreign member of the Royal Society, Britain's prestigious academy of sciences. The timing of this honor coincided with persistent rumors reaching Britain that Vavilov had "somehow fallen out of favor" with the Kremlin and had disappeared. The result was a diplomatic incident.[9]

The Royal Society required the membership diploma to be signed by Vavilov himself and sent it to the Soviet academy president, Vladimir Komarov, in Moscow. But the Soviet academy had been evacuated to Alma-Ata in Kazakhstan. A British embassy diplomat traveled to Alma-Ata to get the signature. The embarrassed Komarov, on instructions from a higher level, presumably, made Nikolai's brother, Sergei, sign the diploma with the Vavilov family name only, hoping this would satisfy the British. Sergei apparently felt that he had no option. But the British diplomat was not so easily fooled, and Komarov received a sharp letter from the embassy. "We expected the signature of Nikolai Vavilov, not Sergei." Evidently, the British had gotten word of the attempted subterfuge.[10] The matter was not resolved.

A second communication to Beria concerned Vavilov's old professor, Dmitry Pryanishnikov. He had worked tirelessly for Vavilov's release, or for a reduction in his sentence. Immediately after Vavilov's arrest, Pryanishnikov, a familiar figure in the halls of the Academy of Sciences and its research institutes, had courageously declared Vavilov's innocence to anyone who would listen, government official, academic, or student. He even obtained an audience with Beria, and petitioned for Vavilov's release. In the spring of 1941, he had written a strong letter to Beria criticizing Lysenko's leadership of the Lenin Academy, and a year later, Pryanishnikov had sent a telegram to Moscow submitting Vavilov's world collection and recent works for the Stalin Prize. It was a crime to promote an enemy of the people and that could have resulted in his own arrest had he not had special connections. Beria's wife was a friend and had worked with Pryanishnikov at the Petrovka.

Pryanishnikov's persistence, and the Royal Society's intervention, finally got Beria's attention. Nikolai Ivanovich and his cell mate Ivan Luppol received word on July 4, 1942, that their death sentences had been commuted to twenty years in a labor camp.[11]

Vavilov and Luppol were transferred from death row to a general cell where they were allowed to exercise and, every so often, have a bath. A local scientist, an entomologist named Megalov, learned that Vavilov was in the prison and sent him food parcels.

The first three parcels were received and a short note, "Thank you, N. Vavilov," came back.[12] Luppol was soon moved out, but Vavilov waited to hear through the autumn and into the winter about his fate. Although his spirits were high his body was wasting away.

In these days, a younger biologist briefly shared a cell with him. On his first day, the biologist saw a gray-haired specter of a man sitting by the window, writing something on a piece of paper with the stub of a pencil. A pile of papers rested on the windowsill nearby. As he drew closer he recognized Vavilov, whom he had met many years earlier.

The young man would recall that he said, "Greetings, Nikolai Ivanovich!"

"Vavilov looked at me intently . . . to my grave I will never forget those mournful eyes."

"You used to know me?" Vavilov asked, slowly.

"Yes, certainly, Nikolai Ivanovich. In 1933 or '34 we were together many times. In those years, when I was still at the Yaroslav station."

After a moment's reflection, Nikolai Ivanovich embraced his former student and invited him to sit beside him. He did not ask what his crime was, nor his sentence. Instead, the two men talked about the Institute, and of lost colleagues. Vavilov related his last trip to the western Ukraine. Then supper arrived. The guards pushed the soup through the feeding slot in the door of the cell.

"I handed Nikolai Ivanovich a bowl, a spoon and a piece of bread," the young man remembered. "He ate very little. In the same bowl I gave him some of my own. He took it and said, 'Remember Misha, a dish, like a man, should be clean inside and out.'

"I apologized and said that they did not give us any water to wash the dishes.

"'I know, old friend, I know . . . ' Vavilov replied."

After the meal, the prison guards took Vavilov away. His younger colleague helped him gather his notes, his bundle of bedding, and his towel.

"With tears in my eyes I shook his gentle hand. I sensed that this was our last handshake."[13]

In their room in Yelena's sister's house a short distance from Prison No. 1, Yuri and his mother were also trying to survive. There was an iron stove that could heat up the room quickly, but as soon as the fire was out, it got cold again. They slept in their coats. The young Yuri was the log cutter, since Maxim, his uncle, had been drafted into a tank regiment. With her paralyzed hands, his mother could do almost nothing. She especially had difficulty darning Yuri's socks.

Yelena applied for employment at the Faculty of Agronomy— where she had met Nikolai Ivanovich in the winter of 1917—but there were no jobs for family members of an enemy of the people. On weekends, Yuri would go to the local factory that pressed sun-flower seeds for oil, and collect the mash to chew for a little suste-nance. The Germans bombed the factory and caused a great fire, the worst attack Saratov had experienced so far during the war. Refu-gees arrived daily from Leningrad with horror stories of the siege, and there were constant rumors about the battle for Stalingrad. The daily bread ration was three hundred grams of black bread per per-son, the same as in the prison. Every so often a letter with money would arrive from Sergei, who had been evacuated with his physics institute to Yoshkar-Ola, five hundred miles to the north.

Sergei also had no idea where Nikolai was in prison, and he found life without his brother "unbearably hard."[14] His dreams would include visions of Nikolai, "skinny, with scars of baked blood." In the spring of 1942 he wrote in his diary, "Still nothing about Nikolai, as if he is dead. Maybe he is dead."

The money he sent to Yelena and Yuri kept them going but also could be a source of envy and trouble. After one payment from Sergei, Yelena sent Yuri to the market with enough money to buy a whole sack of potatoes. With some difficulty he lugged the sack home on his back and dragged it down into the basement of the house. But a neighbor was watching and during the night broke the lock of the basement door and stole the sack. Yelena wept bitterly and Yuri tried to comfort her.[15]

In the middle of February of the winter of 1942–43, the NKVD summoned Yelena to the Saratov headquarters, known as the "Gray House." She went trembling with fear that something terrible was going to happen, that Yuri, for some reason, was about to be arrested. But the agent did not ask about her son. He asked, repeatedly, when she had last seen Nikolai Ivanovich, and whether, and under what circumstances, she had heard from him after he was arrested. She told him, over and over, that she had not seen him since the day he left for Ukraine in July 1940, and she had heard nothing since. They let her go.

The agents knew that Nikolai Ivanovich had died. On January 24, 1943, emaciated and suffering from a fever, he was moved to the prison hospital. As he entered he introduced himself, "You see before you, talking of the past, the Academician Vavilov, but now according to the opinion of the investigators, nothing but dung."[16] In the hospital he complained of chest pains and shortness of breath. He had diarrhea and was hardly eating. He was put on "meals of the 2nd type" that included milk, and glass jars were placed on his chest in an effort to extract the fever that the doctor noted might be a recurrence of malaria.

A committee of Saratov jail doctors examined Nikolai Ivanovich. They noted that he was complaining of overall weakness. And they saw "maceration, pale skin, and swelling in the feet." Their diagnosis was: "dystrophy from prolonged malnutrition."[17]

•   •   •

On January 26, 1943, at seven o'clock in the morning, Nikolai Ivanovich's heart stopped beating. The official cause of death was recorded as pneumonia, a cold caught in the prison exercise yard, perhaps. This eminent plant hunter who had a plan to feed the world had died of starvation.

CHAPTER 29

"Oleg, Where Are You?"

The NKVD notoriously lied about the fate of its victims. Information about arrested persons was designed to keep family members confused, and in fear.

In the case of Nikolai Ivanovich, state security spread a rumor in June 1943 that the famous plant hunter had died in Saratov. The recipient of this rumor was not Yelena or Yuri, who were still living a short walk from the Saratov prison, but Nikolai's brother, Sergei. He arranged for Oleg, Nikolai's son, to go to Saratov to investigate. Oleg was now a strong, handsome young man of twenty-five, and a promising employee of the Lebedev Physical Institute of the Academy of Sciences. Oleg immediately took the train to Saratov. This was not easily done, especially for a relative of an enemy of the people, and required official sanction, but Oleg obtained the necessary travel papers from Sergei.

Oleg's first stop was the house where Yelena and Yuri were staying. He gave no advance warning of his arrival, as any missteps on such a perilous quest might send the wrong signals to agents of state security. Oleg told Yelena and Yuri that information about Nikolai's

280

death might be available at the Gray House, the NKVD headquarters. It was the same building where police had summoned Yelena five months earlier and interrogated her about what she knew of Nikolai Ivanovich. She knew nothing then, and had heard nothing since.[1]

The following morning Oleg and Yuri walked to the Gray House and entered its forbidding reception hall on the first floor. Oleg knocked on the shuttered window, the daunting barrier between common citizens and the state ministry. He insisted that the sixteen-year-old Yuri stay on the other side of the reception room. He did not want him to be involved.

The knock on the window was eventually answered and a hand pulled up the shutter. From where he was, Yuri could see a man's face, but he did not hear the short conversation before the shutter was closed again with a deadening finality. Oleg told Yuri that the official had said any inquiries had to be made at the NKVD headquarters in Moscow. The two brothers returned to Yuri's house, and Oleg announced that he would go back to Moscow immediately. Yuri wanted to accompany him at least to the train station, but Oleg wouldn't allow it.

Oleg did not go back to Moscow immediately. He went to the post office and sent two telegrams, one to his wife, Lidia, and the other to his uncle Sergei. The telegrams read simply, "Died, January 26th, 1943. Send cables to me Saratov, post restante, Oleg."[2] Evidently Oleg had confirmed his father's death, but had spared Yelena and Yuri the awful news.

"A terrible cable," Sergei wrote in his diary. "The cruelest death among my kin. . . . My reaction: to die, by all means, to die. . . . And Nikolai so much wanted to live. . . . God is everything a mistake?"[3]

When Sergei told his wife, Olga, she cried out, with a piercing yell. "I will never forget Olyusha's scream," Sergei wrote in his diary. "My own soul is frozen and become a stone. I work and live as a robot, squeezing myself hard. I long for a quiet, fast, invisible death."[4] A colleague visited Sergei shortly afterward and remembered, "Usually when I called on them we had tea and some discussion, but on this day Olga Mikhailovna was very upset. They had received notice of Nikolai Ivanovich's death and Sergei had gone to his room, from which he eventually emerged. You know Sergei Ivanovich, like Nikolai had a dark complexion, a Russian darkness, black hair and dark skin. They got it from their mother, Alexandra Mikhailovna. But when I saw Sergei, I got a fright, his face was close to black, apparently the ink color of his blood vessels added to his dark skin. No one said a word. He was sitting like a person who is somewhere else, like Orpheus in the Inferno when Eurydice disappeared."[5]

For Sergei, there was no relief. As director of the Physics Institute, he was busy that fall reorganizing the staff as they returned to Moscow. The institute was now turning its attention to the theories of nuclear physics and the power of the atom. It was urgent and absorbing work, but Sergei could not get Nikolai out of his thoughts. When he looked in the mirror he saw Nikolai. "I recognize his gestures and features. I am wearing his overcoat," he wrote in his diary.

Some time later, Sergei was called to the NKVD Moscow headquarters to sign a paper confirming the date of Nikolai's death. "The last thin thread of hope is torn. I should fully get it—Nikolai is dead."[6] Still missing was information about how Nikolai had died, whether he had been shot, or had died from some other cause, and where he was buried.

Oleg was determined to find out and Sergei was willing to help. In October, Sergei arranged for Oleg to make another trip to Saratov. This time Oleg did not contact Yelena and Yuri, and it is not clear when they learned about Nikolai's death. According to Yuri, and he is quite sure about this, Yelena never spoke about Nikolai

Ivanovich's death, either at that time or later.[7] In fact, after the arrest in 1940, Yelena never talked about why or how his father had disappeared. It was as though she was trying to protect him from possessing information that would only make trouble for him.

On this second trip, Oleg discovered that prisoners were buried in Saratov's Voskresensky cemetery, near the prison. During the war, the undertaker was a man named Alexei Novichkov.[8] When he received instructions to collect the dead from the prison, he would pile the naked bodies onto his cart, or onto his sleigh in winter, and dump them into unmarked graves at the cemetery. Novichkov drove metal spikes into the ground where the prisoners' bodies were buried to warn grave diggers the area was occupied. Each body was identified with a metal name tag attached to one leg, and Novichkov never bothered to look at the names. He was only interested in collecting his "payment," a bottle of surgical spirit to wash his hands. But he did not "waste" the spirit on his hands, he recalled later.

In his search, Oleg may have talked to Novichkov, or found the area of the cemetery marked with the metal spikes, but he told only Sergei about his discovery. Yelena and Yuri continued living in Saratov without knowing how near they were to Nikolai's unmarked grave. Sergei sent them money, which was especially appreciated in the winter months. In December Yelena wrote to him, "I cannot express enough deep gratitude to you—without your help we would not have been able to survive in these times."[9] Even so, their life was terribly hard. In the spring of 1944, Yuri planted cucumber seeds in the garden that produced an astonishing bounty that would surely have made his father smile. Yuri fertilized the young plants with horse manure that he picked up in the road from passing carts, and watered them frequently. There were so many cucumbers that he carried them to the covered market and sold them.

In May, Sergei returned to Leningrad after the German siege had ended. The city was devastated and empty. In St. Isaac's Square he stopped at Nikolai Ivanovich's Institute. It was deserted except for a handful of workers who had courageously stayed through the siege to protect the world collection. At least four of them had died of starvation at their desks never giving in to the temptation to eat the precious seeds. A. G. Stchukin, a groundnut specialist, died at his writing table. The keeper of the oat collection and the head of the herb laboratory also succumbed. After the death of a rice specialist, D. S. Ivanov, the staff found several thousand packs of rice that he had preserved.[10] Nikolai's dedicated workers had dug in the center of St. Isaac's Square to plant potatoes from the collection, a necessary routine to keep the valuable seed reservoir alive. "If only God and souls could be brought back inside these walls," Sergei wrote in his diary.[11]

Sergei crossed the square, past the Astoria Hotel where Hitler had planned the German victory celebrations, and walked down Gogol Street to Nevsky Prospekt. It was evening and there was a light on in Nikolai's apartment, a single lamp with an orange shade shining onto the street below. He had expected the apartment to belong to someone else. Enemies of the people lost everything. (Nikolai's apartment had been given to a young ballerina, Natalia Dudinskaya.) Sergei knew that there was worse to come: Nikolai's name, his whole life's work, would be omitted or crossed out from scientific reference books. The pain was all too vivid as Sergei saw how easily the society was now sweeping his brother out of sight.

With the end of the war in Europe, in May 1945, an increasing number of Nikolai's international colleagues, especially in Britain and America, began asking about him. In June, scores of foreign scientists attended the celebrations in Moscow marking the 220th anniversary of the Academy of Sciences. They learned that Nikolai had died sometime between 1941 and 1943, but there were no details. The British Royal Society repeatedly asked the

Academy of Sciences for information, but never received a reply.

In July 1945, Stalin struck again at the Vavilov family. He appointed Sergei Vavilov president of the USSR Academy of Sciences. Sergei was undoubtedly qualified. He had an international reputation as a physicist in the field of light, fluorescence, and optics. Yet the appointment was an act of extreme callousness: one immediate result was to deflect international criticism of Nikolai's arrest and death away from the Kremlin, and from Stalin personally. And Sergei was put in an impossible position: to refuse Stalin's call might have been to court disaster, and yet to accept was to do the bidding of the man who had murdered his brother. The compliant members of the Academy and the NKVD provided a suitable cover for the appointment.

The former president, Vladimir Komarov, had retired and Stalin ordered the academy to submit a list of prospective candidates for the job. The academy submitted twenty-two names, including Sergei Vavilov and Trofim Lysenko. Each candidate had been assessed by state security. Sergei was the brother of Nikolai, but he was highly qualified, according to the NKVD. He was at the height of his scientific powers, he had great organizing abilities, his manner was "straightforward, his way of life modest"—and he was "politically loyal." This was in contrast to the other candidates, one of whom was said to be a "heavy drinker," another was "quarrelsome and reserved," and yet another was unsuitable because he "did not associate with his colleagues due to the excessive snobbery of his wife." Lysenko, on the other hand, had a major fault. He did not "enjoy respect," including that of the outgoing president, Komarov. And, "everyone blames him for N. I. Vavilov's arrest."[12]

Sergei's file did not mention something that was, perhaps, his most important qualification. He was a physicist with administrative experience, and the most urgent task of Soviet scientists in 1945 was to build an atomic bomb. (Inevitably, Sergei was involved in the Soviet bomb project—and he would joke, "There used to be two methods to find the truth: induction and deduction; now there is a third, 'Information.'" He meant espionage.)[13]

In the Soviet Union, for a moment, Vavilov's appointment might have seemed a blow against Lysenko—even, perhaps, that the era of repression of geneticists might be over. But Stalin had in no way recanted his Lamarckian beliefs or his choice of Lysenko as the leader of "Soviet biology."

Some of Sergei's colleagues were disgusted by his decision to serve Stalin. They could not understand how he could deliver fawning speeches praising Lysenko's neo-Lamarckism when his brother had become a martyr to genetics. For his own part, Sergei had doubts that he could do the job. He would have to meet regularly with Lysenko, for one thing. For another, he would have to pretend that his brother's many contributions to Soviet science never existed. He rationalized his decision by telling himself that he was serving science, not Stalin—a defense similar to the one Nikolai had made through the dark days of the 1930s, that he was serving science and Mother Russia, not the Communist Party.

But postwar Russia was briefly a different place for the bourgeois specialists. Scientists had made key contributions to the war effort—and were about to make an even more significant contribution to producing the atomic bomb. The power of the party functionaries and the *vydvizhentsy* was in temporary decline. With the prestige of science now high in the Kremlin, the lot of members of the Academy of Sciences would improve significantly—with higher salaries and privileges, including their own dachas.

Sergei Vavilov would play an important role in this resurgence. He insisted on importing foreign laboratory equipment and again giving academy members access to foreign scientific publications—both of which had been severely restricted. Whenever possible he also helped colleagues in need with cash from his own pocket. And he was now in a more powerful position to help Nikolai Ivanovich's sons, Oleg and Yuri, as well as Yelena, although he had never really approved of her.

A few days after his appointment, Sergei arranged for Yuri and his mother to return to Leningrad. In Saratov, Yuri received a telegram addressed to "the employee of the academy of sciences of the Soviet Union, Yuri Nikolayevich Vavilov, who should proceed immediately to Leningrad together with his mother Yelena Barulina and start work at his new place of employment."[14] It was signed by the vice president of the Academy of Sciences, Comrade General Leon Orbeli, a physiologist and supporter of the geneticists. Of course, there was no job. Yuri was only seventeen and still in high school. It was a ploy organized by Sergei to get Yelena and Yuri to Leningrad.

They went first by train to Moscow where they met Sergei in the Moskva Hotel, near Red Square. As the new president of the academy Sergei was given a two-story mansion in the fashionable Arbat, but the house was being renovated. At the hotel, Yelena spoke with Sergei for about thirty minutes—while Yuri stayed in another room. It is possible that was when Sergei told Yelena about Nikolai's death and the details of his burial that Oleg had discovered. It is also possible that he never mentioned it. The main thrust of the conversation was about the move back to Leningrad and that their old apartment on Gogol and Nevsky Prospekt was now occupied by a ballerina.

Sergei told Yelena that he had arranged for accommodation in the academy's apartments on Vasilevskaya Island, across the Neva. The apartment was small, one big room, and they built a bookcase down the middle of the room to divide it into "equal shares." In June 1946, Yuri graduated from high school, winning a silver medal. The award allowed him to enter Leningrad University without further exams, and without any mention of him being an "enemy of the people." With Uncle Sergei's help, the younger of Nikolai Ivanovich's two sons was now established in Stalin's Russia.

•   •   •

But Sergei was powerless when it came to protecting Oleg. In late January 1946, three years almost to the day after Nikolai Ivanovich's death, Oleg left Moscow for a two-week skiing trip to Dombai in the northern Caucasus. He was a good skier and this trip was to celebrate the end of his doctoral studies at Moscow University. His wife, Lidia, did not want him to go, but Oleg insisted. He had not had a holiday for several years. "I need to unwind," he told her.

She recalled the night of his departure. Oleg did not own an appropriate coat or a backpack, and a friend loaned him a German camouflage jacket and German mountain boots—the spoils of war. As he was about to leave, his mother, Katya, came into the kitchen to say goodbye. He hugged his mother and Lidia and promised to return soon.

Two weeks later, on February 10, Lidia received a call from the Science Sport Society, which had arranged the trip. She was told there had been an accident. Oleg had fallen to his death. There had been a heavy snowfall, and they had been unable to find his body. When Lidia told Katya, she cried out loud, blaming herself for letting him go, and called out, "Oleg, where are you? Oleg, where are you? My hopes were futile, I shouldn't have cherished hopes for you, then I would have avoided drinking from this cup, the same as for Nikolai Ivanovich."[15]

Lidia went to Sergei for help and he agreed to finance a ten-member search party, including some of the Science Sport Society's most experienced climbers. The guide, a tall, stout man named Schneider, who had been the last person to see Oleg alive, was in the party. He introduced himself as a historian but, oddly, no one could find out where he worked. When they arrived on the scene Schneider was strangely reluctant to help in the search. After ten days the group reached the area where Oleg was supposed to have disappeared. The snow was up to eighteen feet deep in places. They found nothing and returned home.

Back in Moscow, a rumor spread, possibly from the NKVD, that

Oleg had escaped to Turkey. But Lidia refused to believe it, and insisted on launching a second search. On June 17, Lidia returned to the site with another search party, again paid for by Sergei. Even in summer the snow was still thick. As Lidia probed it with her ice pick, she saw some red clothing. She cleared the snow frantically with her hands, brushing away enough to see Oleg's red-checked shirt. As she and her companions dug, Oleg's body was slowly exposed.

"I saw him as if he was alive," she would recall. "Dark hair, thick eyebrows, even his cheeks were red. His pockets still contained his passport." She found his backpack and his ice pick nearby. The search party carried his body down the mountain and the local police made out a death certificate and gave her a copy. The document stated that there was a "wound in the area of the right temple the size of an ice pick's adze."[16] Oleg's body was put into a sealed coffin and buried on the spot.

Back in Moscow, the Science Sport Society held a meeting to discuss his death. There was a stormy debate during which it was discovered that one of the original leaders of Oleg's skiing group had talked his comrades out of looking for him.[17] Was his death an accident, or was he murdered by a blow from an ice pick? Like so many mysterious deaths in this terrible period there was no way to know. The Vavilov family, however, would have no doubts about why Oleg died. He knew too much about his father's death; he was becoming a prominent member of the Soviet scientific community and would have an opportunity to tell the truth to his colleagues, not only in the Soviet Union but also to visitors from abroad. Stalin did not want the full story of Nikolai Vavilov ever to be told.

Vavilov's Ghost

Since threats forced Galileo's historic recantation, there have been many attempts to suppress or distort scientific truth on behalf of various creeds alien to science, but not one of these attempts was long successful.

—SIR HENRY DALE, PAST PRESIDENT OF THE ROYAL SOCIETY, 1948

In the summer of 1946, Russia suffered its worst drought in half a century. In the southern steppes where the soil was rich and black, the wheat harvest was half its prewar level.[1] The country was plunged into its third famine since the 1917 revolution. In Ukraine, where Nikita Khrushchev was then first secretary of the republic's Communist Party, collective farmers begged for help. They had given their grain quota to the state, as required, and there was nothing left.

That was when the cannibalism started, Khrushchev would write in his memoirs. "I received a report that a human head and the soles of the feet had been found under a little bridge near Vasilkovov, a town outside Kiev. Apparently, a corpse had been eaten."[2] It is not known how many thousands, or millions, died of starvation or how many turned to cannibalism. The newspapers censored reports of the famine, and the tragedy would not be known in the West until Khrushchev, who led the country after Stalin's death in 1953, published his memoirs almost twenty years later. The drought came amid the postwar devastation of Soviet agriculture, with its catastrophic loss of livestock, harvesting machinery, and manpower.[3] In addition, food was already being used as a weapon in the new Cold War. As Russians starved, Stalin diverted tons of Soviet grain to France—hoping to influence the outcome of the first postwar election to the French parliament.

There was little that any plant breeding methods, of whatever kind, could have done to forestall this part natural, part man-made disaster, but Russian geneticists had already planned their return. The time seemed right. Geneticists had lost their leader in Nikolai Ivanovich and many of his prominent researchers, but the spirit and science of Vavilov continued to be felt at home and especially abroad. During the war, the scientific institutes had been evacuated from Leningrad and Moscow, and those geneticists who had escaped arrest continued to work quietly. When the war ended a few of them began to speak out about the need to restore genetics to its rightful place in Soviet biology.

They wrote letters to the press criticizing Lysenko, especially his latest idea of heredity that "every drop of protoplasm [of the cell, not just the nucleus, as geneticists believed] possesses heredity."[4] The geneticists also tried to oust Lysenko from key posts. Importantly, the Central Committee's science department became so concerned about Lysenko's monopoly on agricultural progress that it ordered a review of the work of his Lenin Academy of Agriculture. Officials demanded new critical appraisals of the "Lysenko school," complaining that Lysenko had "held back" Soviet agriculture by his outright rejection of hybrid corn, which was proving so successful in America, and he had failed to keep his promises to produce new improved varieties of crop plants in two to three years.[5]

The geneticists also sought help from their Western colleagues. Relations between the Western and Soviet scientists had improved markedly during the anti-Hitler alliance, reviving the idea of a "single-world science." American scientists, still unsure of the fate of their friend Vavilov, wrote articles praising his achievements in plant breeding and his unrivaled world collection of seeds. An article in *Science* noted that the Russians had "outstripped the Germans in this field even before the advent of Hitler."[6]

Vavilov's international standing would receive a surprise boost when the Allies discovered that Hitler had created a special SS commando unit to loot the seeds in Vavilov's collections. As the unit had advanced, many small seed banks in western Russia were captured. But the Germans missed the prize at Vavilov's Leningrad Institute because they were never able to take the city.[7]

As scientists outside Russia slowly learned that Vavilov had died in jail, the obituaries started to appear. In Britain, he was remembered in the journal *Nature* as an outstanding botanist who had "in practice laid the foundations of all future improvements in crop plants."[8] The authors, Vavilov's close friend, the British cotton breeder S. C. Harland, and the geneticist Cyril Darlington wrote that Vavilov's potato collection "led to the establishment of the British Empire Potato Collection on which potato breeding is now being based in Britain and elsewhere."

They recalled the "remarkable sight" of watching Vavilov at work in his Institute in Leningrad "in his shirt sleeves, sprawled over a map of the Soviet Union covering the floor of his office, busy distributing and arranging his staff and stations. They concluded, "A host of friends in Europe and America will lament his death . . . but science at large will remember his achievement, an achievement that survives his personal disaster."

In Russia, Stalin at first approved of the warming trend in the international science community. The war had given Stalin a different view of "bourgeois" scientists. The American success in building an atomic bomb was stark evidence that science did not depend on social class, but on expert training, money, and resources. The Soviet slogan became "catch up and overtake" the West. And this was not only in atomic physics.

Stalin announced new state assistance for all Soviet scientists. In a famous declaration in 1946, he said, "I have no doubt that if we give our scientists the help they need, they will soon catch up and even surpass the achievements of science abroad."[9] The next month's national budget for science tripled and the salaries of scientists increased. The emphasis was on building a Soviet atomic bomb, but other departments gained support as well. For a while, it seemed to outsiders as though the geneticists might be making a comeback. "It is safe to assume that Lysenko's school has passed its zenith," wrote Eric Ashby, an Australian botanist who spent a year in Russia after the war as a member of his country's diplomatic mission.[10]

The opportunity for a resurgence of genetics would be short-lived, however. Two factors intervened. The swift onset of the Cold War—"the final establishment of two opposing camps, Soviet and Western,"[11] as the Russian historian of science, Nikolai Krementsov, put it, and Stalin's stubborn adherence to Lamarckism, his enduring belief that "trained" miracle varieties of crops could solve the problems of Soviet agriculture. In Stalin's Kremlin office a bizarre scene was unfolding.

On December 30, 1946, Stalin summoned Lysenko to the Kremlin, but this time it was not about "bourgeois" scientists "wrecking" the revolution, or even about the postwar offensive by Soviet geneticists. It was to discuss the potential for Soviet agriculture of a peculiar type of wheat, known as branched wheat. Branched wheat produces more than one spike and, thus, it would seem, more grain.[12]

Stalin, the amateur gardener proud of the lemons and other fruits he grew in a hothouse at his dacha outside Moscow, had received

samples of branched wheat from a loyal farmer in his native Georgia. Apparently believing that this wheat could make a serious difference to desperate Soviet grain shortages, Stalin ordered Lysenko
to start planting the variety on Russia's collective farms.

Lysenko carried out the order immediately and with his customary enthusiasm. He organized a mass planting and he appeared in
newspaper photographs clutching sheaves of branched wheat as the
variety was promoted in the press as a new wonder crop that could
bring food to the masses.

The effort was a failure, as any reader of Russian scientific literature could have predicted. Branched wheat had been tried a hundred
years earlier and rejected. It required a particularly rich soil, and lots
of room because it produced many leaves as well as spikes. Most
important of all, bread could not be made from its flour because it
did not contain enough gluten.

Such matters were of little concern to the obsequious Lysenko,
who promised Stalin more and greater plantings. And Stalin
rewarded Lysenko with his praise, reassuring him that his theory of
plant breeding was "the only correct scientific approach" and that
the geneticists who "deny the inheritance of acquired characteristics
do not deserve to be discussed." Stalin was so enthusiastic about
branched wheat that he sent copies of Lysenko's report to all
members of the Politburo with a letter, marked "Strictly Secret,"
directing the matter to be raised for further discussion.[13]

Lysenko used this praise to claim an intimacy with Stalin that
did not in fact exist. As Zhores Medvedev has pointed out, Lysenko
never met Stalin except officially, and he was never invited to
Stalin's dacha to see his fabled hothouse lemons. But at this crucial
time in the evolution of Soviet biology, Lysenko's official contacts
with Stalin, their shared belief in Lamarckism and the inheritance
of acquired characteristics, would propel Lysenko to a final victory
over the geneticists and establish him as the virtual dictator of Soviet
biology.[14]

• • •

Lysenko complained to Stalin that he could no longer work efficiently under the postwar wave of attacks on his theories and offered to resign as president of the Lenin Academy. Stalin ordered the Central Committee to end its criticism and orchestrated an elaborate plan to squash the geneticists once and for all. In the summer of 1948, the Politburo hastily convened one last "public discussion" of the Lenin Academy, now with additional members approved by Stalin and heavily rigged in Lysenko's favor. Lysenko would give the main address, which had been edited by Stalin himself.

This was no casual glance at Lysenko's speech. Stalin's pencil marks were all over the original draft. The Berlin crisis was at its height; the world had shifted and so had Stalin's argument. He changed the title of the conference to "The Situation in Biological Science," a worldview, not a local view as promised in Lysenko's original title of "Soviet" biological science. Stalin took out references to "bourgeois" as opposed to "proletarian" science, and turned it into a contest between "idealistic" (Mendel) and "materialistic" (Lysenko) theories. When Lysenko made a reference to all science being class-based, Stalin crossed it out. He added in the margin his usual taunt to anything he found unacceptable, "Ha, ha, ha!" Then he wrote, "And [what about] mathematics?" And he added emphasis to Lysenko's declaration of the absolute truth of Lamarckism, even attaching an entirely new paragraph for Lysenko to reassert the supremacy of Lamarck's theory. "It cannot be denied that in the controversy that flared up between the Weismannists [critics of Lamarck who believed inheritance was affected only by the germ cells] and Lamarckians in the beginning of the twentieth century, the Lamarckians were closer to the truth for they defended the interests of science, whereas the Weismannists were at loggerheads with science and prone to indulge in mysticism."[15]

The Lenin Academy met between July 31 and August 7, 1948. Lysenko's speakers, dogmatic, insulting, and dismissive as before, outnumbered the geneticists seven to one. The geneticists called for an "end to the polemics and the enmity," but it was not to be. Stalin was determined that this session should end the dispute— in Lysenko's favor. On the evening of August 6, the night before the last day of the conference, Stalin again called Lysenko to his Kremlin office. He told Lysenko to announce that his direction in biology—Lysenkoism—had been approved by the Central Committee, something that had not yet occurred. Stalin then dictated the first paragraph of Lysenko's concluding remarks.

On the morning of August 7, 1948, Lysenko wound up the conference by declaring, "Comrades, before I pass to my concluding remarks, I consider it my duty to make the following statement. The question is asked in one of the notes handed to me, 'What is the attitude of the Central Committee of the party to my report?' I answer: The Central Committee of the Party examined my report and approved it." In the record of the meeting, the stenographer added, "Stormy applause. Ovation. All rise." Sergei Vavilov would write in his diary, "The newspapers are full of Lysenko. It's his triumph."[16]

Twenty-one years after the skinny youth from Karlovka, Ukraine, had been picked out of the wheat fields by the *Pravda* correspondent as the poster boy of the up-and-coming Soviet specialist in agriculture, Trofim Lysenko had been crowned the king of biology. It was, Stephen Jay Gould, would later write, "the most chilling passage in all the literature of twentieth century science."[17] In the eight days of so-called public discussion, which filled 631 pages of transcript, there was not a single mention of Nikolai Vavilov, his Institute in Leningrad, the experimental farm in Pushkin, or his famous world collection of seeds. It was as though they had never existed.

• • •

In the purge that followed about three thousand biologists lost their jobs to Lysenko's followers in universities, institutes, and schools of higher education across the country. The director of the old Petrovka, the agricultural academy, was replaced and Lysenko himself grabbed the academy's key department of crop breeding and seed production. Many others were subjected to harassment and humiliation and required to denounce their views. Researchers were hounded out of Vavilov's old Institute.

The geneticist Vladimir Efroimson, a longtime critic of Lysenko who had been arrested in 1932, freed in 1935, and had joined in the postwar criticism of Lysenko, was rearrested as a "socially dangerous element."[18] Dmitry Sabinin, a plant physiologist, was expelled from Moscow State University but then repeatedly persecuted and finally shot himself. Yosif Rappaport, a geneticist and war veteran who had been a severe critic of Lysenko at the Lenin Academy meeting in August, was expelled from the party.

Lysenkoists celebrated their triumph with scientific degrees, honorary titles, prizes, medals, and orders; they were given salary increases, apartments, dachas, and personal cars. Lysenko was awarded another Order of Lenin to mark his fiftieth year. The citation was "for outstanding services in the cause of developing advanced science and great, fruitful practical activity in agriculture."[19]

When news of the July session of the Lenin Academy reached friends and admirers of Nikolai Ivanovich around the world, there was outrage. Sir Henry Dale, who had been president of Britain's Royal Society until 1945, resigned his honorary membership in the U.S.S.R Academy of Sciences. (He submitted his resignation to the Soviet academy's president, Sergei Vavilov.) Sir Henry noted that Nikolai Vavilov, a man he described as "this outstanding Russian scientist," had been replaced by Lyensko, "a proponent of a doctrine of evolution that in essence contradicts all the success achieved by investigators in this field since the time in the early 19th century when Lamarck's arguments were first published." He charged that the "great structure of precise knowledge which is continuing to grow by the efforts of the heirs of Mendel, Bateson and Morgan is being subjected to denial and abuse." He called the Soviet Academy's resolution to promote Lysenkoism in Soviet schools and universities "a clear expression of this political tyranny."

Lysenko's forced victory turned into his own state-sponsored cult. His portrait hung in all scientific institutions, and art stores sold busts and bas-reliefs. In some cities monuments were erected. Sergei Vavilov watched the destruction of his brother's legacy and his own subservience in increasing pain and his health began to suffer. In 1949, Sergei wrote to Stalin pleading for Nikolai's rehabilitation. In the handwritten letter, Sergei described his childhood with Nikolai, what a big influence his brother had been on his life, how much he had helped him, and how he would not have been a great scientist without this help.

The letter described Nikolai's work in some detail, his openness and the straightforwardness of his scientific judgments. Sergei categorically denied that Nikolai had been involved in any hostile acts, and pleaded for his brother's pardon. "If my brother is not rehabilitated, I cannot be the president of the Academy of Sciences," he wrote. The letter went first to Beria's desk where the secret police chief stamped it "Not authorized."

There is no evidence that the letter ever reached Stalin, but on July 14, 1949, Stalin received Sergei for a talk that lasted an hour and a half. Sergei's diary entry says it was "about the Academy and the [Soviet] Encyclopedia. He [Stalin] met me rather strictly without a smile . . . he was smiling on my way out. . . . [He complained that] the Academy is fooling around and not producing anything. . . . I submitted fifteen papers. . . . In general, I don't know if it's good or bad. I'm full of anxiety. I must take a holiday, everything is so unclear."[20]

· · ·

During 1950, Sergei made a special effort to help Yelena Barulina and Yuri, who were still living in one room in Leningrad. He arranged for them to move into a two-room flat on Baskov Lane—it was unheard of for the family of an "enemy of the people" to get such an apartment. The NKVD was still playing its cruel game, however. Yelena was called to the Registry Office in Leningrad, supposedly to receive Nikolai's death certificate. When she opened it, Nikolai's name was on the certificate but the details of his death had been left blank.[21]

On January 25, 1951, Sergei died of a heart attack. He was given an elaborate funeral, as befitted a high government official. Yuri Vavilov remembered that the coffin "sunk in flowers" as it lay in the House of the Union in Moscow and people came from all over the Soviet Union to pay their respects.[22]

Lysenko's absolute domination of Soviet agriculture lasted for six years, but the geneticists never gave up and in 1952 the first mildly critical articles since the 1948 session of the Lenin Academy accused Lysenko of retreating from Darwinism and not properly representing Lysenkoism's "authority," Ivan Michurin.[23] As mild as that seems now, the articles unleashed a flood of more critical reports. In 1953, Stalin died, aged seventy-four, and Khrushchev took over, but it would be another twelve years before Lysenko was finally ousted from his key posts.

In 1955, more than three hundred scientists signed a petition requesting Lysenko's removal as president of the Lenin Academy. Abroad, the American corn breeder Paul Mangelsdorf had by then added his voice in a Vavilov obituary. "It is quite fortunate for us that the free world should now begin to reap substantial harvests, both scientific and utilitarian, from the very work which Vavilov's own country disdained as 'impractical.'" Lysenko fought back, portraying all criticism as a Western imperialist plot and appealing for protection against the slander. But that same year, Nikolai Ivanovich Vavilov was "rehabilitated."

The Central Committee opened a review of cases under Stalin involving the death penalty. A number of scientists, including Lysenko, were interviewed. Now without his protector, Stalin, Lysenko acted as though he had never uttered a bad word against Vavilov. He characterized him as "a big Soviet scientist, an organizer of science" and one whose "big connections of global samples of seeds of various agricultural plants deserves the highest positive evaluation." Lysenko said he disagreed with the charges of sabotage against Vavilov; he was "not destroying collections, on the contrary, he was creating them." He concluded, "All other analogous charges on topics of science, in my view, are not founded either."[24]

The Central Committee brought suspect evidence in Vavilov's trial to the attention of the Military Collegium of the Supreme Court of the USSR. The chief military prosecutor, Major Justice Koleshnikov, requested Vavilov's conviction to be rescinded and the case terminated. On August 20, 1955, military judges reviewed the evidence, agreed with Koleshnikov, and issued Judgment No. 4 n-011514/55. In a brief summary, they noted that in the original trial in 1941, Vavilov only partially admitted guilt, and after the conviction declared that he had "never engaged in any counterrevolutionary activity."

The judges charged that "gross breaches" of the Procedural Code were committed during the preliminary investigation of Vavilov's case and that the investigation was "tendentious and unobjective." The most glaring example was the testimony of a dead man, Alexander Muralov, on August 7, 1940, the day after Nikolai's arrest. Muralov had replaced Vavilov as president of the Lenin Academy in 1935, but he was arrested and shot in 1937, three years before he was alleged to have given evidence. The special prosecutor's report declared, "This shows that case material against Vavilov was falsified." It named Vavilov's interrogator, Alexander Khvat, as a known "falsifier of investigative material."

Yelena Barulina was the first to receive the judgment. One can only imagine her relief—and her anger—after fifteen years of torment and terrible pain for her loss. Despite her weakened condition, she had been waiting for this moment to prepare for publication of Nikolai Ivanovich's manuscripts, especially his *World Resources of Cereals and Flax and Their Application for Selection*. She collaborated with Nikolai's young researcher Fatikh Bakhteyev, who had accompanied Nikolai on his last expedition to Ukraine in 1940. *World Resources* was published in 1957 by the Academy of Science publishers. One month later, Yelena Barulina died, weakened by her crippling nerve disease. She was sixty-two years old. Nikolai's divorced wife, Yekaterina Sakharova, continued to live in Moscow and, strangely, she was working in a branch of the KGB's library. She died in 1964, aged seventy-eight.

Under Khrushchev, Lysenko managed to strengthen his position, only losing power after Khrushchev himself was ousted in 1964. Lysenko even survived the unraveling of the genetic code by James Watson and Francis Crick and the beginning of the new era of molecular genetics. Lysenko was finally forced to resign as director of the Institute of Genetics in 1965, a move that effectively ended his reign. His resignation did not appear in the press, although it was quickly known. Two years later, in 1967, Vavilov's Institute, which, since 1930, had borne the anonymous, cumbersome title of the All-Union Institute of Plant Industry, was renamed the N. I. Vavilov Institute of Plant Industry. Lysenko was allowed to continue his work on the effects on plants of temperature, humidity, and other environmental factors. When he was blamed, as he sometimes was by colleagues, for the death of Nikolai Ivanovich, he would explode in anger, "I did not kill Vavilov!"

The KGB chairman, Vladimir Semichastny, continued to protect Lysenko. He sent a letter to Khrushchev about the "puffed up" rumors "from Dmitry Pryanishnikov and others" that Lysenko had been involved in Vavilov's death. The letter concluded, "The archive material contains no data to confirm them."[25] The long list of Lysenko's extravagant promises went unfulfilled—vernalization, new varieties of spring wheat, the renewal of self-fertilized plants, the summer planting of potatoes and sugar beets, the planting of winter wheat on Siberian stubble, and branched wheat. But he never surrendered.

In 1971, the American historian of science Loren Graham met Lysenko by chance in the cafeteria of the Academy of Sciences in Moscow. Lysenko started to berate Graham for "several serious mistakes in describing me and my work." The most important mistake, Lysenko insisted, was accusing him of responsibility for Vavilov's death. Graham remained silent for a while, and then defended his interpretation of Lysenko's actions that led to Vavilov's arrest. Lysenko abruptly got up and left the table. After about ten minutes he returned and launched into a rant about his disadvantaged life. "I came from a simple peasant family," he began. "I encountered the prejudices of the upper classes. Vavilov came from a wealthy family and knew many foreign languages . . . most of the prominent geneticists of the 1920s were like Vavilov. They did not want to make room for a simple peasant like me."[26] Lysenko never acknowledged the key role Vavilov had played in the recognition of his early work. He was an active member of the Academy of Sciences, but he sat alone in the cafeteria "ostracized," as he put it, by all the other scientists. He died in 1976.

By the middle seventies, not only had Nikolai Vavilov's work been restored to the institutes and libraries but he had become an icon of Soviet science, "a new saint for the orthodox," in the words of one visiting American plant breeder.[27] In 1987, the Russians celebrated the centenary of Nikolai Vavilov's birth with a flurry of articles in newspapers and magazines. It was the period of Mikhail Gorbachev's glasnost. An article in the mass circulation magazine *Ogonyok* said that Vavilov's "struggle for truth against all the baseness, demagoguery and arbitrariness that were part of the cult of personality is particularly precious today in the light of our rethinking of past and present and our quest for a worthy future."[28]

In the *Moscow News* weekly, the investigative journalist Yevgeniya Albats published an interview with Vavilov's interrogator, Alexander Khvat. He told her, "I never believed the spying charges, of course, there was no evidence. As for sabotage he was doing something in a strange direction in science." Albats asked, "Didn't you pity Vavilov . . . he was going to be shot?" To which Khvat replied, "Ah, there were so many people!"[29]

Speeches at the Lenin Academy talked about the "sad fact" of the 1948 session. The president, Alexander Nikonov, said that he was now "authorized to say" that the academy's current staff "condemns, rejects and refuses to acknowledge Lysenkoism." A note about the celebration and Vavilov's "rehabilitation" from the British embassy to the Foreign Office in London ended, "One more blank spot has been filled."[30] By then, Vavilov's entry in the *Dictionary of Scientific Biography* read, "In the West and the Soviet Union, Vavilov has come to be regarded as one of the outstanding geneticists of the twentieth century, a symbol of the best aspects of Soviet science, and a martyr for scientific truth."[31]

As for his scientific achievements, Vavilov was always more concerned with the application of genetic principles to the problems of agriculture than with developing new theories of his own. During his last days in Saratov in 1920, he demonstrated considerable skills of self-promotion in launching his Law of Homologous Series— especially in suggesting that it was the biological equivalent of Mendeleev's periodic table. The favorable press at the time encouraged this comparison and certainly helped to establish him as a scientific authority in Petrograd when he took over the Bureau of Applied

Botany in 1921. But in those days the way genetic information was transferred from genes into characteristics was not yet known. Vavilov's law would come under criticism as the action of genes became clearer. In 1937, Vavilov himself admitted, "We underestimated the variability of the genes themselves. . . . At the time we thought that the genes possessed by close species were identical; now we know that this is far from the case, that even very closely related species that have extremely similar traits are characterized by many different genes."

Today, scientists have modified and updated Vavilov's original centers of origin of cultivated plants; they debate how many such centers actually qualify as "original," and whether they should be renamed, in the modern nomenclature, "centers of biodiversity." Jack Harlan, the son of Vavilov's American plant breeding friend, Harry Harlan, became a leading plant explorer and breeder. He refined Vavilov's eight centers into two types, a true center, where the plant originated, and a "noncenter," where crops were domesticated by farmers, such as Africa for wheat, or South America in the case of corn. Harlan credited Vavilov with three "bull's-eyes"—his centers of Peru for potatoes, Oaxaca (Mexico) for corn, and Palestine for wheat. "By and large," Harlan concluded, Vavilov's 1926 essay "Centers of Origin of Cultivated Plants" was a landmark and is still influential. "As of that date it was a remarkable perception, but based more on intuition than data."[32]

As Harlan concluded in 1995, "The world of N. I. Vavilov is vanishing and the sources of genetic variability he knew are drying up. The patterns of variation [that Vavilov described on his expeditions] may no longer be discernible in a few decades and living traces of the long coevolution of cultivated plants may well disappear forever."

Vavilov's world collection is his enduring legacy. The seed bank was never a dead herbarium—it was a living museum of cultivated plants, lovingly cared for by loyal and devoted researchers—and it is still widely acclaimed. The seeds in his Leningrad, once again St. Petersburg, collection are still in demand—by those who would protect and preserve the genetic diversity of the world's food plants from catastrophe, natural or man-made. A global seed bank is no longer the dream of one man in a former tsarist palace in St. Isaac's Square, but a twenty-first-century quest by conservationists in a doomsday vault on a Norwegian island, eight hundred miles from the North Pole.[33] Today's latest concept is not a copy of Vavilov's utopian idea of a common bowl from which everyone can feed, however. The world, and genetics, have moved on. Now, 7 percent of the earth's arable land is sown with genetically modified crops, new varieties heavily patented by agribusiness that has no plans to share them. And, for the moment, each nation will have access only to its own seeds. Even so, these seeds will hopefully be safe from anti-science demagogues, ideology, censorship, willful neglect, and political expediency.

Main Events in the Life of Nikolai Ivanovich Vavilov

1887   Born in Moscow, November 25 (November 13 old style), to Ivan Ilyich Vavilov and Alexandra Mikhailova Vavilova.

1905   First Russian revolution ends in December when tsarist guards enter Presnya, the industrial suburb of Moscow where the Vavilov family lived. Death of Nikolai's seven-year-old brother, Ilya, of appendicitis.

1906   Graduated from Emperor's High School of Commerce.

1911   Graduated from Petrovskaya Agricultural Institute, known as the "Petrovka."

1911–12   Intern with Robert Regel at the Bureau of Applied Botany of the Scientific Committee of the Ministry of Land Cultivation in St. Petersburg.

1912   Married Yekaterina Nikolayevna Sakharova.

1912–14   Studied in Britain, France, and Germany, especially with William Bateson at the John Innes Horticultural Institute at Merton, near London. First World War begins in 1914. Nikolai's younger brother, Sergei, is drafted into the army. Nikolai avoids the draft because of an eye injury. Nikolai's younger sister, Lydia, dies of smallpox, October 18, 1914.

1916   Expedition to northern Persia for the Imperial Army, and to the Pamirs for the Ministry of Agriculture.

1917   February, Tsar Nicholas II abdicates. October, Bolsheviks seize power.

1917–20   Vavilov is a professor at the Faculty of Agronomy, University of Saratov. Russian civil war begins, 1918. Son Oleg born, November 7, 1918.

1920   Formulated the Law of Homologous Series.

1921–40   Director, Bureau of Applied Botany, Petrograd (Leningrad in 1924), and of successor institutions, Department of Applied Botany and New Crops (1924–29), All-Union Institute of Plant Industry, 1930–40. [In 1967, the institute was renamed the All-Union N. I. Vavilov Institute of Plant Industry (VIR).]

1921   First visit to United States—to Washington, D.C., in connection with American aid for the Russian famine. Visits U.S. Department of Agriculture and experiment stations across United States and Canada.

1922   Elected corresponding member of Academy of Sciences of the USSR.

1924   Expedition to Afghanistan. Death of Lenin, Stalin takes over.

1926   Awarded the Przhevalsky Gold Medal by the Russian Geographical Society for his Afghanistan expedition. The first among five scientists to be awarded the Lenin Prize. Divorced Yekaterina Sakharova and became common-law husband of Yelena Barulina.

1926–27   Expedition to Abyssinia (now Ethiopia). Also visited

North Africa, the Middle East, and Mediterranean countries. Awarded the Gold Medal of the International Conference on Agriculture in Rome, 1927. Publication of Vavilov's *Studies on the Origin of Cultivated Plants*.

1927   First public appearance of Trofim Lysenko in an article in *Pravda*. Vavilov biography in the first edition of the *Great Soviet Encyclopedia*. Vavilov's father, Ivan Ilyich, dies.

1928   Second son, Yuri, born, February 6, 1928.

1929   Stalin's Year of the Great Break with the Past. Vavilov is elected a full member of the Academy of Sciences of the USSR. Visits northwest China, Japan, Formosa, and Korea. Confirmed by the USSR Council of People's Commissars to be the president of All-Union Lenin Academy of Agriculture. Vavilov organizes the first Soviet Congress on genetics and plant breeding.

1930–40   Director of the Institute of Genetics of the Academy of Sciences of the USSR.

1930   Expedition to Central America and Mexico.

1931   Soviet delegate, with Nikolai Bukharin, to the 2nd International Congress on the History of Science and Technology, London. Elected president of the Geographical Society of the USSR.

1932   Vice president of the 6th International Congress of Genetics, Ithaca, New York.

1932–33   Last expedition abroad—to North and South America: United States, Canada, Mexico, Central America, Colombia, Peru, Bolivia, Chile, Argentina, Uruguay, Brazil, Trinidad, Puerto Rico, and Cuba.

1934   Soviet government cancels the celebration in Leningrad of the tenth anniversary of the Institute of Plant Industry.

1935   Stalin praises Trofim Lysenko. And Lysenko launches an offensive against the geneticists.

1936   Showdown with Lysenko. Arrests of geneticists. 7th International Congress of Genetics, scheduled for Moscow, canceled by the government.

1938   Death of Vavilov's mother, Alexandra, April 5, 1938.

1939   Vavilov is elected honorary president of the 7th International Congress of Genetics in Edinburgh, Scotland, but not allowed to attend. Molotov–Ribbentrop nonaggression pact between USSR and Germany. World War II begins.

1940   Death of Vavilov's sister Alexandra, April 3, 1940. Vavilov arrested August 6, 1940, during an expedition in Ukraine.

1940–41   Interrogated by NKVD.

1941   June 22, Germany invades the Soviet Union. Vavilov found guilty of sabotage of Soviet agriculture and spying for Britain. Sentenced to be shot, July 9, 1941. Sentence commuted to twenty years in corrective labor camps.

1942   Elected foreign member of Royal Society, London.

1943   January 26, Vavilov dies in Saratov prison hospital, aged fifty-five, of malnutrition.

1945   Stalin appoints Sergei Vavilov president of the Academy of Sciences of the USSR.

1946   Oleg Vavilov dies in mysterious skiing accident in the Caucasus.

1948   Lysenkoism becomes the official direction of Soviet biology.

1951   Publication in the United States in *Chronica Botanica* of *Selected Writings of N. I. Vavilov. The Origin, Variation, Immunity* *and Breeding of Cultivated Plants.* Vavilov's biography removed from the second edition of the *Great Soviet Encyclopedia*. Death of Sergei Vavilov, January 25, 1951.

1953   Stalin dies, Khrushchev takes over.

1955   Vavilov rehabilitated. Yelena Barulina begins work on his manuscripts, especially *World Resources of Cereals, Leguminous Seed Crops, and Their Utilization in Plant Breeding.*

1957   Death of Yelena Barulina, July 9, 1957.

1962   Posthumous publication of *Five Continents: Tales of Exploration in Search of New Plants*, from incomplete manuscript.

1964   Death of Yekaterina Sakharova.

1987   Centenary of Vavilov's birth celebrated in Moscow.

Vavilov was honorary member of Moscow Society of Investigators of Nature, the Linnean Society of London, New York Geographic Society, Botanical Society of America, Mexican Society of Agronomy, and Natural Sciences Society of Spain, as well as an honorary doctor of Sofia University, Bulgaria; foreign member of the British Royal Society; honorary member of Indian Academy of Sciences, member of Royal Scottish Academy, foreign member of Czechoslovak Academy of Agricultural Sciences, corresponding member of Deutsche Akademie der Naturforscher Leopoldina in Halle.

Sources and Archives

The basic Russian language sources for biographical material on Nikolai Vavilov, his family, and his confrontation with Trofim Lysenko are to be found in the following archives, books, and journals (some of which have been translated into English):

1. Vavilov's official papers that survived searches by agents of state security immediately after his arrest in 1940. These papers are at Vavilov's institute, Vsesoyuzny Institut Rastenievodstva (VIR—All-Union Institute of Plant Industry, now named for Vavilov), St. Petersburg. A collection of his correspondence and scientific papers is also in Moscow at the archive of N. I. Vavilov's Memorial Study, N. I. Vavilov Institute of General Genetics, Russian Academy of Sciences, Moscow. The international correspondence is in six bound volumes, Nikolai Ivanovich Vavilov, *Nauchnoe Nasledie v Pismah*, Mezhdunarodnaya Perepiska [*Scientific Heritage in Letters, International Correspondence*], ed. R. V. Petrov, V. K. Shumny, and A. A. Zhuchenko (Moscow: Nauka, 1994–2003). The edition covers the period 1921–40 and consists of 3,963 letters, 1,404 to foreign correspondents and 2,559 letters received by Vavilov from abroad. His official domestic letters are in two volumes, *Nauchnoe Nasledstvo* [*Scientific Inheritance*], Vol. 5, 1911–28, 428 pages, and Vol. 10, 1929–40, 491 pages (Moscow: Nauka, 1987). N. I. Vavilov, *Ocherki, Vospominaniya, Materiali* [*Sketches, Memoirs, and Materials*] (Moscow: Nauka, 1987). Vavilov's account of his expeditions, *Piat Kontinentov* [*Five Continents*] (Moscow: Mysl, 1987), was put together from an incomplete manuscript preserved by A. S. Mishina, who from 1938 to 1940 worked as a typist and stenographer for Nikolai Vavilov. She hid the manuscript during the war and evacuation of Leningrad. The manuscript has been translated into English by Doris Love, edited by Semyon Reznik and Paul Stapleton, and published by the International Plant Genetic

Resources Institute (IPGRI), Rome, 1997. A second key publication of IPGRI in English is *Vavilov and His Institute: A History of the World Collection of Plant Genetic Resources in Russia*, by Igor Loskutov, Rome, 1999.

2. The personal archive of Nikolai's son Yuri Vavilov, in Moscow, contains family letters, documents, and photographs. The diaries of Nikolai Vavilov at the Petrovka agricultural academy are in N. I. Vavilov's Studenchesky Dnevnik [Student Diary], 1907–11, Russian Academy of Sciences journal, *Chelovek* [*The Human Being*] (Moscow: Nauka, 2005), nos. 2, 3 and 5. Sergei Vavilov's diaries are in *Voprosi Istorii Estestvoznaniya y Techniki* [*Matters of History of Natural Sciences and Technology*] (Moscow: Nauka, 2004), nos. 1 and 2; (no. 1 covers 1909–16); (no. 2 covers 1939–51); S. I. Vavilov's diaries for 1947 are published by Novaya Gazeta, *Kentavr* (*Centaurus*) scientific popular supplement, issue no. 3, Moscow 2007; for 1949, issue no. 4, 2007. Also, S. I. Vavilov, *Ocherki y Vospominaniya* [*Sketches and Memoirs*] (Moscow: Issdatelstvo Nauka, 1979).

3. Several biographies and reminiscences begin in 1932. A. Roskin, *Karavani, Dorogi, Kolosya* [*Caravans, Roads, and Ears of Grain*] (Moscow: Molodaya Gvardia, 1932; Zhores Medvedev, whose samizdat manuscript, *The Rise and Fall of T. D. Lysenko*, was circulated in Russia in the mid-1960s and published in the United States by Columbia University Press, 1969; Semyon Reznik, *Nikolai Vavilov* (Moscow: Molodaya Gvardia, 1968); Mark Popovsky, "Tysiacha dnei Akademika Vavilova" [A Thousand Days of Academician Vavilov], *Prostor*, Alma-Ata, nos. 7 and 8, 1966. This article was the basis for a biography of Nikolai Vavilov in English by Mark Popovsky entitled *The Vavilov Affair* (Hamden, Connecticut: Archon Books, 1984); Fatikh Bakhteyev, *Nikolai Ivanovich Vavilov, 1887–1943* (Novosibirsk: Nauka, Sibirskoe Otdelenye, 1987); Soratniki Nikolaya Ivanovicha Vavilova, Issledovately Genofonda Rasteniy [Associates of Nikolai Ivanovich, Recollections of Researchers of Plant Gene Pool], St. Petersburg, 1994; E. N. Sinskaya, *Vospominaniya o N. I. Vavilove* [*Recollections of N. I. Vavilov*] (Kiev: Naukova Dumka, 1991); E. S. Levina, *Vavilov, Lysenko, Timofeyev-Ressovsky, Biology in the USSR, History and Historiography* (Moscow: Association of Researchers of Russian Society, 1995); Yakov Rokityansky, Yury Vavilov, and V. Goncharov, *Sud Palacha, Nikolai Vavilov v Zastenkak NKVD*) [*Hangman's Court*] (Moscow: Akademiya, 2000); Yakov Rokityansky, "Syn Genya" [Genius's Son], *Chelovek* [Russian Academy of Sciences Journal], no. 3/2001 and no. 1/2004, Nauka; Yuri Vavilov, *V Dolgom Poiski* [*Long Search*] (Moscow: FIAN, 2004); Vladimir Esakov, biographi-

cal sketch, "Put, Kotory Vibirayu" [The Path I Chose], *Chelovek*, 2005, nos. 5 and 6; Vladimir Esakov, *Akademiya Nauk v Resheniyah Politburo CK KPSS 1922–1952* (Moscow: Rospen, 2000); I. V. Shaikin, *Nikolai Vavilov* (Moscow: Molodaya Gvardia, 2006).

4. Papers and documents scattered among several official Soviet government and scientific archives, including:

Moscow: Rossiiskii Gosudarstvennyi Arkhiv Ekonomiki (RGAE) [Russian State Archive of the Economy].

Rossiiskii Gosudarstvennyi Arkhiv Sotsial'no-Politicheskoi Istorii (RGASPI) [Russian State Archive of Sociopolitical History].

Arkhiv Rossiiskii Akademii Nauk (ARA) [Archive of the Russian Academy of Sciences].

Rossiiskii Gosudarstvennyi Arkhiv Noveishei Istorii (RGANI) [Russian State Archive of Contemporary History].

St. Petersburg: Tsentral'nyi Gosudarstvennyi Archiv Nauchno-Tekhnicheskoi Dokumentatsii Sankt-Petersburg (TsGANTD SPb) [Central State Archive of the Scientific-Technical Documentation of St. Petersburg].

St. Petersburg Branch of the Archive of the Russian Academy of Sciences (PFA RAN).

Archive of the N.I. Vavilov Institute of Plant Industry (VIR), St. Petersburg.

5. The basic English language sources, excluding those mentioned above, are: Julian Huxley, *Heredity, East and West, Lysenko and World Science* (New York: Schuman, 1949); David Joravsky, *The Lysenko Affair* (Cambridge: Harvard University Press, 1970); Mark Adams, "Vavilov, Nikolai Ivanovich," *Dictionary of Scientific Biography*, vol. 15, Suppl. 1 (New York: Charles Scribner's Sons, 1978), pp. 506–13; Barry Mendel Cohen, "Nikolai Vavilov, His Life and Work," dissertation, University of Texas, 1980; Loren Graham, *Science, Philosophy, and Human Behavior in the Soviet Union* (New York: Columbia University Press, 1987); Loren Graham, *Science in Russia and the Soviet Union, A Short History* (Cambridge: Cambridge University Press, 1993); Valery Soyfer, *Lysenko and the Tragedy of Soviet Science*, trans. Leo Gruliow and Rebecca Gruliow (New Brunswick: Rutgers University Press, 1994); Nikolai Krementsov, *Stalinist Science* (Princeton: Princeton University Press, 1997); Tanya Lassan, "The Bureau of Applied Botany," *Journal of the Swedish Seed*

Association 107, 1997; Vadim Birstein, *The Perversion of Knowledge: The True Story of Soviet Science* (Boulder: Westview, 2001); Zhores Medvedev and Roy Medvedev, *The Unknown Stalin*, trans. Ellen Dahrendorf (London: I. B. Taurus, 2003); Nils Roll-Hansen, *The Lysenko Effect* (New York: Humanity Books, 2005).

U.S. archives: U.S. Department of Agriculture; Hoover Institution Archives, Stanford, California.

U.K. archives: The National Archives of the United Kingdom, Kew, London.

Notes

EPIGRAPH

Aleksandr Solzhenitsyn, *The Gulag Archipelago*, vol. 2 (New York: Harper & Row, 1974), p. 314.

INTRODUCTION

1. L. Graham, *Moscow Stories* (Bloomington: Indiana University Press, 2006).
2. Zhores Medvedev, *The Rise and Fall of T. D. Lysenko* (New York: Columbia, 1969) p. 58.

PROLOGUE: UKRAINE, AUGUST 1940

1. Vavilov's arrest is taken from accounts in Fatikh Bakhteyev, *Nikolai Ivanovich Vavilov* (Novosibirsk: Nauka, 1987), pp. 216–18; Medvedev, *Rise and Fall of T. D. Lysenko*, pp. 69–70; Mark Popovsky, *The Vavilov Affair* (Hamden, Connecticut: Archon, 1984), pp. 127–28; Barry Mendel Cohen, "Nikolai Ivanovich Vavilov, His Life and Work" (Ph.D. diss., University of Texas, 1980), pp. 227–30.

CHAPTER 1: MOSCOW, DECEMBER 1905

1. The account of Nikolai Vavilov coming under fire during the siege of Presnya is based on S. I. Vavilov in N. I. Vavilov, *Ocherki, Vospominaniya, Materiali* (Moscow: Nauka, 1987), pp. 90–91; A. N. Ipatiev, N. I. Vavilov,

Ocherki, pp. 90–107; and Leon Trotsky, *1905*, trans. Anya Bostock (New York: Vintage, 1972), pp. 234–49.

2. Leon Trotsky, *1905*, p. 245.

3. V. D. Esakov, "Put, Kotory Vibirayu," *Chelovek*, 2005, no. 5, pp. 152–68. Ivan Ilych was a director of the trading company N. and A. Udalov Bros. and I. Vavilov Trade House, founded January 1, 1890.

4. L. Trotsky, *1905*, p. 5.

5. Trotsky's phrase, ibid.

6. *Provintsialnie Khlopoti* no. 2, 2002, p. 97, Ivashkovo.

7. V. Esakov, "Put, Kotory," *Chelovek*, vol. 6, 2005, pp. 152–68, 149–68.

8. Ibid.

9. Barry Mendel Cohen, "Nikolai Ivanovich Vavilov, His Life and Work," diss., University of Texas, December 1980.

10. Semyon Reznik, *Nikolai Vavilov* (Moscow: Molodaya Gvardia, 1968), pp. 9–10.

11. Ibid.

12. Ibid.

13. Diane Koenker, *Moscow Workers and the 1917 Revolution* (Princeton: Princeton University Press, 1981), pp. 54–55.

14. *The Golden Book of Moscow Business, The Year 2000*, joint stock company Trekhgornaya Manufaktura.

15. In fact, the Emperor's High School of Commerce had a good selection of science courses—physics and chemistry, but not biology. Many years later Sergei Vavilov would call it a "great secondary school." Quoted in I. V. Shaikin, *Nikolai Vavilov* (Moscow: Molodaya Gvardia, 2006), p. 12.

16. Ibid.

17. A. N. Sokolovsky in N. I. Vavilov, *Ocherki*, p. 110.

CHAPTER 2: THE PETROVKA AND KATYA

1. L. Breslavets, "He Put Science Above Everything Else," in N. I. Vavilov, *Ocherki*, p. 85.

2. Nikolai Krementsov, *Stalinist Science* (Princeton: Princeton University Press, 1997), p. 14.

3. N. I. Vavilov, Diaries, 1907–11, *Chelovek*, Vol. 5, pp. 138–51.

4. Krementsov, *Stalinist Science*, p. 19.

5. Breslavets, "He Put Science," p. 88.

6. Ibid., p. 87.

7. N. I. Vavilov, Diaries, October 22, 1907, December 31, 1908.

8. Nils Roll-Hansen, *The Lysenko Effect* (New York: Humanity Books, 2005), p. 32.

9. N. I. Vavilov, Diaries, October 22, 1907; December 31, 1908.

10. Krementsov, *Stalinist Science*, pp. 20, 55. Before the revolution, Nikolai Koltsov had created the privately funded Institute for Experimental Biology and this was greatly expanded with state funds after 1917. Yuri Filipchenko and Nikolai Koltsov included materials on genetics in their courses on experimental zoology at Petrograd and Moscow universities.

11. An exception had been a liberal professor Kliment Timiryazev, who was a teenager when Darwin published the *Origin of Species* in 1859. He was among the first Russian botanists to publish in favor of Darwin's theory. A superb popularizer of science, he became the most influential biologist in Russia at the end of the nineteenth century as he started to write about Darwin and Mendel. He was not a geneticist and never studied heredity experimentally. His attitude toward Mendel's laws has been described as "superficial and subjective." He respected Mendel but was skeptical of the total importance attached to Mendel's laws of heredity by the enthusiastic followers of genetics. He had been unable, like Darwin, to come to a firm conclusion on Lamarck's idea of the inheritance of acquired characteristics. He accepted that acquired characteristics could be inherited through grafting. He believed that Mendelian and Lamarckian assumptions were not mutually exclusive. His liberal politics were too much for the tsarist authorities and he was fired as a professor at the Petrovka before Nikolai Ivanovich arrived and he moved to Moscow University. Toward the end of his life—he died in 1920—the liberal Timiryazev saw opposition to Darwin as the efforts of the clerics to confound the theory of evolution and he suggested that the supporters of Mendel were "endeavouring to obscure the importance of Darwinism with the over-inflated glory of a Catholic monk." Such statements would endear him to the Bolsheviks, who would co-opt his name, after his death, for their campaign against Soviet geneticists and would rename the Petrovka academy after him. See more on Timiryazev in Medvedev, *The Rise and Fall of T. D. Lysenko*, p. 29; and David Joravsky, *The Lysenko Affair* (Cambridge: Harvard University Press, 1970), p. 205.

12. See V. Esakov, "Put, Kotory"; K. I. Pangalo in *Ryadom s Vavilovim* (Moscow: Sovietskaya Rossia, 1973), p. 83; also M. Popovsky, *The Vavilov Affair*, p. 20.

13. N. I. Vavilov, Student Diaries, December 8, 1910.

14. Ibid., April 2, 1911.

15. Y. N. Sakharova, private collection, 1904–59. RGAE, Rossiiskii Gosudar-stvenny Arkiv i Ekonomiki, f. 328, d. 37.

16. N. I. Vavilov to Y. N. Sakharova, quoted in Popovsky, *The Vavilov Affair*, p. 22.

17. Ibid.

18. See Zhores Medvedev, *Soviet Agriculture* (New York: W. W. Norton, 1987), pp. 7–18.

19. VIR archives, N. I. Vavilov to Regel, October 18, 1911; also published in *Nauchnoe Nasledstvo*, Moscow, 1980, Vol. 5, p. 18.

20. R. Regel to N. I. Vavilov, October 28, 1911, VIR archives.

21. Popovsky, *The Vavilov Affair*, p. 22.

22. Ibid., p. 23.

23. A.Yu. Tupikova, *Ryadom s N. I. Vavilovim, Sbornik Vospominaniya*, 2nd edition (Moscow: Sovietskaya Rossia, 1973), p. 33. Students at one of his seminars surprised him with a celebration. "We poured a rain of confetti on him from the balcony and then warmly congratulated him. He was very embarrassed, and also very moved."

CHAPTER 3: IN DARWIN'S LIBRARY

1. N. I. Vavilov, "Darwin on the Origin of Cultivated Plants," Session on Darwin, Soviet Academy of Sciences, November 28, 1939, author's notes, *Soviet Science*, no. 2, 1940, pp 55–75.

2. Ibid.

3. See R. Pistorius and J. van Wijk, *The Exploitation of Plant Genetic Information: Political Strategies in Crop Development* (Wallingford, Oxon, UK: CABI Publishing, 1999), p. 36.

4. William Bateson, *Problems of Genetics* (New Haven: Yale University Press, 1979), p. 16.

5. William Bateson, address to the British Association at Melbourne, 1914; also in ibid., p. xvi.

6. N. I. Vavilov, "William Bateson, 1861–1926. In Memory of a Teacher," *Trudi pro prikladnoi botanike i selektsii*, 1926, Vol. 15, no. 5, pp. 499–511.

7. Yakov Rokityansky, "Genius's Son," *Chelovek 6* (2003): pp. 111–30.

CHAPTER 4: MOSCOW, SUMMER 1916

1. The account of Nikolai Vavilov's departure for the Persian front is based on the recollections of his nephew, A. N. Ipatiev, in N. I. Vavilov, *Ocherki*, p. 91; also on S. I. Vavilov diaries in *Voprosi Istorii Estestvoznaniya y Techniki* [Matters of History of Natural Sciences and Technology] (Moscow: Nauka), 204, nos. 1 and 2.
2. Sergei Vavilov, Diaries, June 16, 1914.
3. Ibid., December 6, 1914.
4. Ibid., May 25, 1916.
5. Lidia Vasilievna Kurnosova, quoted in Rokityansky, "Genius's Son," *Chelovek*, no. 6/2003, pp. 111–30.
6. Popovsky, *Vavilov Affair*, p. 14.
7. R. Regel to the Council of Heads of the Agricultural Scientific Committee, October 12, 1917, TsGANTD SPb, f. 179, op. 2, d. 103, 1.4–6.
8. Recollections of A. N. Ipatiev in N. I. Vavilov, *Ocherki*, p. 90.
9. Ibid.

CHAPTER 5: ON THE ROOF OF THE WORLD, 1916

1. According to Vavilov's report of his trip, the Demri-Shaurg glacier is "within Fergana." N. I. Vavilov, *Five Continents*, ed. L. E. Rodin, Semyon Reznik, and Paul Stapleton; trans. Doris Love (Rome: International Plant Genetic Resources Institute, 1997), p. 9.
2. The quotations from Vavilov's travels are in ibid., pp. 5–16.
3. Ibid., p. 9.

CHAPTER 6: REVOLUTION AND CIVIL WAR

1. Sergei Vavilov, Diaries, May 25, 1916.
2. Cohen, "Nikolai Ivanovich Vavilov," p. 66.
3. Ibid., p. 63.
4. Popovsky, *Vavilov Affair*, p. 25.
5. Donald Raleigh, *Experiencing Russia's Civil War: Politics, Society, and Revolutionary Culture in Saratov, 1917–1920* (Princeton: Princeton University Press, 2002), pp. 6–10.
6. Ibid., p. 278.
7. Ibid., pp. 275–78.

8. Ibid., p. 251.

9. N.I. Vavilov, "The Problems of Origin, Geography, Genetics, Plant Breeding, Plant Industry and Agronomy," *Selected Works* [Russian], vol. 5 (Moscow: USSR Academy of Sciences Press, 1965), p. 440.

10. Popovsky, *Vavilov Affair*, p. 26.

11. N. I. Vavilov to R. Regel from Saratov, September 14, 1919, Central Archive of Scientific-Technical Documentation of St. Petersburg (TsGANDT SPb, fond 179, op. 1–2, d. 639, 1.292-293.

12. E. P. Podyapolskaya-Ramenskaya, in N.I. Vavilov, *Ocherki*, p. 149.

13. Ibid.

14. N. I. Vavilov to R. Regel, *Nauchnoe Nasledstvo*, Vol. 5, 1911–1928 (Moscow: Nauka, 1980), p. 34.

15. Reznik, *Nikolai Vavilov*, p. 108.

16. N.I. Vavilov to R. Regel, November 3, 1918, *Nauchnoe Nasledstvo*, vol. 5, p. 36.

17. A. N. Ipatiev, in N. I. Vavilov, *Ocherki*, pp. 91–107.

18. K. I. Pangalo, "Science Turns into Art," in N. I. Vavilov, *Ocherki*, p. 162.

19. R. Regel, People's Commissariat of Agriculture certificates, TsGANDT SPb, 179-1/2—639-208.

20. N. I. Vavilov to Regel, September 29, 1917. N. I. Vavilov, *Nauchnoe Nasledstvo*, 1911–28, Vol. 5, pp. 27–29.

21. Robert Regel to N. I. Vavilov, November 22, 1917, Archive of the N. I. Vavilov All-Russian Institute of Plant Industry, fond/opis 1, d.4, 1.23–24ob.

22. N. I. Vavilov to Regel, December 11, 1917.

23. Robert Regel to N. I. Vavilov, November 22, 1917, archive of N. I. Vavilov All-Russian Institute of Plant Industry, fond 1, d. 4, 1, 23–24ob. Regel was worried about the Germans reaching Petrograd; on March 4, 1918, German troops stood within a hundred miles of the city.

24. N. I. Vavilov to Regel, November 3, 1918. *Nauchnoe Nasledstvo*, vol. 5, p. 36.

25. N. I. Vavilov to Regel, September 14, 1919, Central Archive of Scientific-Technical Documentation of St. Petersburg (TsGANDT SPb) fond 179, op. 1–2, d. 639, 1 292–293.

26. N. I. Vavilov, *Nauchnoe Nasledstvo*, Vol. 10, p. 399.

27. M. E. Sergeenko, in N. I. Vavilov, *Ocherki*, p. 141.

28. N. I. Vavilov to Regel, May 11, 1919. TsGANDT SPb, fond 179, op. 1–2, d. 639, 1.235.

29. Regel recommended Vavilov in a letter to the council of heads of departments of the Agricultural Scientific Committee, October 12, 1917. TsGANTD SPb, f. 179, op. 2, d. 103, 1.4–6.

30. Popovsky, *Vavilov Affair*, p. 26.

31. Although hailed as a biological Mendeleev, Vavilov's Law of Homologous Series would not stand the test of time in the same way. In 1975, the director of the Harvard Botanical Museum, Dr. R. E. Schultes, noted, "After a century this law is still actively discussed among geneticists and plant breeders. Today, the parallelism in variation patterns in plants and animals is studied with modern biogenetical methods and described with precision exceeding that possible during Vavilov's time. It turns out that plants and animals vary in similar ways when subjected to similar [environmental] pressures." "Vavilov's Law," *Economic Botany* 29 (Oct.–Dec. 1975), p. 378. Also, Vavilov's law did not take into account the variations caused by chemical or radiation-induced mutations.

32. Valery Soyfer, *Lysenko and the Tragedy of Soviet Science* (New Brunswick: Rutgers University Press, 1994), p. 46.

CHAPTER 7: THE GARDENER OF KOZLOV

1. Quoted in Joravsky, *The Lysenko Affair*, p. 48; *Polnoe Sobranie*, vol. 2, pp. 298–300.

2. Joravsky, *The Lysenko Affair*, p. 43.

3. Frank Meyer, letter from Kozlov to David Fairchild, Agricultural Explorer in Charge, USDA, Washington, D.C., December 29, 1911.

4. N. I. Vavilov, "Exploit," *Pravda*, June 8, 1935.

5. Popovsky, *The Vavilov Affair*, p. 44.

6. Gali Popova, *Ryadom s N.I. Vavilovim, Sbornik Vospominaniya*, 2nd ed. (Moscow: Sovietskaya Rossia, 1973), p. 58–60. Compiled by Yuri Vavilov.

7. Ibid.

8. N. I. Vavilov to L. S. Berg, *Nauchnoe Nasledstvo*, July 17, 1980, p. 41.

CHAPTER 8: LENOCHKA

1. Donald Raleigh, *A Russian Civil War Diary: Alexis Babine in Saratov, 1917–1922* (Durham: Duke University Press, 1988), p. 168.

2. N. I. Vavilov to Yelena Ivanovna Barulina, from Moscow to Saratov, October 1, 1920, in Yuri Vavilov, *Long Search* (Moscow: FIAN, 2004), pp. 42–43.

3. Ibid.

4. Cable, R. Regel to N. I. Vavilov, February 8, 1918, VIR archive, fond/opis 1. d. 4, 1.96–96ob.

5. Popovsky, *Vavilov Affair*, p. 32.

6. Reznik, *Nikolai Vavilov*, p. 131.

7. N. I. Vavilov to Yelena Barulina from Petrograd to Saratov, November 27, 1920. Private Archive, Yuri Vavilov.

8. Ibid.

9. Ibid.

CHAPTER 9: PETROGRAD: CITY OF RAVENS

1. Solomon Volkov, *St. Petersburg: A Cultural History* (New York: Simon & Schuster, 1995), pp. 224–25.

2. Ibid., p. 211.

3. Ibid., p. 225.

4. Vadim Birstein, *The Perversion of Knowledge* (Boulder, CO: Westview Press, 2001), p. 35, from *Lenin's Collected Works*, Vol. 44 (Moscow: Progress Publishers, 1970), p. 284, in a letter from Petrograd dated September 15, 1919.

5. Zhores Medvedev, *Soviet Science* (New York: Norton, 1978), p. 6.

6. Gorky and Prokofiev returned when Lenin's New Economic Policy improved life for a while.

7. Krementsov, *Stalinist Science*, p. 18.

8. Popovsky, *Vavilov Affair*, p. 32.

9. Professors were slowly coming back. One of them was Lev Berg, the geography professor, who had taken his family to Ukraine during the civil war. The train back to Petrograd had taken eight days just to cover the four hundred miles from Moscow. Although the Bergs were relatively well off, the family could not find any means of transport from the station and walked the two miles to their new apartment, in the gardener's quarters, not the main house. His daughter Raissa, then five years old, would later write, "There was absolutely nothing to eat. Nanny traveled to the outlying village and exchanged her skirts, which she had got under the tsarist regime, for potatoes and milk." See Raissa Berg, *Acquired Traits: Mem-*

oirs of a Geneticist from the Soviet Union, trans. David Lowe (New York: Viking, 1988), Chapter 1.

10. N. I. Vavilov, *Nauchnoe Nasledstvo,* Vol. 5, 1911–28, p. 41.

11. Ibid.

12. Popovsky, *Vavilov Affair,* p. 32.

13. N. I. Vavilov to W. Bateson, *Nauchnoe Nasledie v Pismah,* Mezhdunarodnaya Perepiska [International Correspondence] (Moscow: Nauka, 1994), vol. 1, Petrograd Period, 1921–27, October 5, 1922, p. 56, no. 42.

14. Berg, *Acquired Traits,* p. 13.

CHAPTER 10: INGOTS OF PLATINUM

1. L. C. Dunn, "Soviet Research in Biological Sciences," *Science in Soviet Russia,* National Council of American-Soviet Friendship (Lancaster, Pennsylvania: Jaques Cattell Press, 1944), p. 28.

2. Krementsov, *Stalinist Science,* p. 22.

3. N. I. Vavilov to E. I. Barulina, June 28, 1921, private archive of Yuri Vavilov.

4. Christine White, *British and American Commercial Relations with Russia, 1918–1924* (Durham: University of North Carolina Press, 1992), p. 161. Between May 1921 and January 1922, $57 million worth of Russian gold passed through the U.S. consulate in Paris for shipment to New York.

5. Ibid., p. 158–59.

6. N. I. Vavilov to Yelena Barulina, June 26, 1921, private archive of Yuri Vavilov.

7. N. I. Vavilov to E. I. Barulina, June 28, 1921.

8. Ibid.

9. H. H. Fisher, *The Famine in Soviet Russia 1919–23: The Operation of the American Relief Administration* (New York: Macmillan, 1927). Also in Medvedev, *Soviet Science,* p. 41.

10. Quoted in I. Loskutov, *Vavilov and His Institute: A History of the World Collection of Plant Genetic Resources in Russia* (Rome: IPGRI, 1999), p. 19, from Vavilov, *Five Continents* [Russian], p. 79.

11. Reznik, *Nikolai Vavilov,* p. 137.

12. Constitution of Association for Promotion of American-Russian Agriculture Inc., TsGANTD SPb, fond 318, op. 1–1, d. 6, 155.

13. P. S. Everest, agent, United States Indian Service, Department of the

Interior, to D. N. Borodin, December 31, 1921, TsGANDT, SPb, fond 318, op. 1–1, d. 6, 1, 86.

14. N. I. Vavilov in J. F. Crow, "Perspectives, Plant Breeding Giants: Burbank, the Artist, Vavilov, the Scientist," *Genetics*, no. 158 (August 2001), p. 1394.

15. Ibid. But Vavilov persisted and as they chatted for several hours, the two men found much in common in their approach to plant breeding, despite Burbank's downplaying of Mendel. Typically, Nikolai Ivanovich wanted to put Burbank into some botanical pigeonhole, to distill his work in order to make sense of it in the wider context of plant breeding. So he listed three principles that seemed to govern Burbank's approach.

   The first was his "mobilization of the earth's plant capital: South American flora, Tibetan and Himalayan mountain species, plants from China, Japan and the entire Old World"—just as the U.S. Department of Agriculture was doing, and as Vavilov himself planned to do at his Petrograd institute.

   Second, Vavilov noticed Burbank's use of seeds from fruit, such as apples and pears, that had produced asexually. By growing plants from seedlings, Burbank was able to cross-pollinate them and create new varieties, with new combinations of genes and new characteristics.

   Third, Burbank created new species by breeding different ones and then using the normal asexual reproduction of fruit trees to fix the new variety in future generations. Vavilov would write in an obituary, "the mind of the ingenious plant breeder grasped, with American quickness, the practicality of this new path of asexual plant reproduction." Burbank created his famous potato from a seedling that he brought to Santa Rosa in 1870 and by 1906 it was the favored potato along America's Pacific Coast.

16. Ibid.

17. D. N. Borodin to N. I. Vavilov, November 17, 1921, TsGANDT SPb, fond 318, op. 1–1, d. 6, 139.

18. N. I. Vavilov, *Nauchnoe Nasledtsvo*, Vol. 5, p. 43.

19. N. I. Vavilov to D. N. Borodin, June 3, 1922, *International Correspondence*, p. 35.

20. N. I. Vavilov to H. Humphrey, *International Correspondence*, vol. 1, May 10, 1923; N. I. Vavilov to D. N. Borodin, New York, ibid., July 9, 1923, p. 81, and November 30, 1923, p. 92; N. I. Vavilov to J. Vilmorin, ibid., March 15, 1924, p. 108; N. I. Vavilov to T. Morgan, ibid., February 20, 1924, p. 104; N. I. Vavilov to W. Bateson, May 12, 1924, p. 113.

21. N. I. Vavilov, *Nauchnoe Nasledtsvo*, Vol. 5, p. 47.

22. N. I. Vavilov, Letters to Bateson, 1922, Archives of John Innes Center. Also quoted in Igor Loskutov, *Vavilov and His Institute*, p. 24.

CHAPTER 11: AFGHANISTAN, 1924

1. *British Foreign Office* files N3637, 28 April 1924, Public Record Office, Kew, London.
2. The quotations from Vavilov's travels are in Vavilov, *Five Continents*, pp. 23–50.
3. See Peter Hopkirk, *The Great Game: The Struggle for Empire in Central Asia* (Tokyo: Kodansha International, 1992).
4. Popovsky, *Vavilov Affair*, p. 69.
5. N. I. Vavilov, *Five Continents*, pp. 34–42. The crops of Nuristan turned out to be extremely poor and no endemic forms were found. The soils were poor and in need of manure. Vavilov postulated many hypotheses about the origin of the Kafirs, including that they are the remnants of the army of Alexander the Great, but he concluded that they seemed to be exiles, driven by fate into impassable forested massifs and inaccessible mountain ravines.
6. N. I. Vavilov to P. P. Podyapolsky, January 13, 1925, *Nauchnoe Nasledstvo*, Vol. 10, p. 171.
7. Z. S. Vedeneeva, in N. I. Vavilov, *Ocherki*, p. 343.
8. N. I. Vavilov to P. P. Podyapolsky, January 13, 1925, *Nauchnoe Nasledstvo*, vol. 10, p. 171.
9. Soyfer, *Lysenko and the Tragedy of Soviet Science*, p. 48, quoting Mark Popovsky, *Nado Speshit* [We Must Hurry] (Moscow: Detskaya Literatura, 1968), p. 97.
10. N. I. Vavilov to M. Gaisinsky, *International Correspondence*, vol. 1, p. 178.
11. K. I. Pangalo, in N. I. Vavilov, *Ocherki*, p. 191.
12. N. I. Vavilov to G. Karpechenko, April 12, 1926, *International Correspondence*, vol. 1, no. 10, p. 160.
13. N. I. Vavilov to P. P. Podyapolsky, January 13, 1925, ibid., vol. 1, no. 226.
14. N. I. Vavilov to P. P. Zhukovsky, May 5, 1924, ibid., vol. 1, no. 218.
15. N. I. Vavilov to S. Bukasov, November 2, 1925, ibid.
16. N. I. Vavilov to M. I. Topolyanskaya, February 20, 1925, ibid., vol. 1, p. 117; N. I. Vavilov to J. H. Reisner, ibid., May 20, 1926, p. 168; N. I. Vavilov to D. A. Borodin, ibid., January 16, 1925, p. 115, and March 4, 1925, p. 117.

CHAPTER 12: ABYSSINIA, 1926

1. N. I. Vavilov to Nikolai Gorbunov, April 3, 1926, *Nauchnoe Nasledstvo*, vol. 5, no. 384.
2. Unless indicated otherwise, the quotations from Vavilov's travels are in Vavilov, *Five Continents*, pp. 81–106.
3. N. I. Vavilov to V. Pisarev, February 17, 1927, *Nauchnoe Nasledstvo*, vol. 5, p. 295.
4. N. I. Vavilov to V. E. Pisarev, April 8, 1927, *Nauchnoe Nasledstvo*, Vol. 5, p. 295.
5. N. I. Vavilov, *Five Continents*, pp. 115–16: "Among the ruins of Herculaneum and Pompeii I was able to see the remnants of agricultural crops dating from thousands of years ago. There were both wheat and barley and the large-seeded flax much the same as in our time. Two thousand years have, in essence, hardly changed the composition of the crops of this ancient, agricultural country. As is well known, the great naturalist at the beginning of our era, Pliny the Elder, perished while studying the igneous rocks at the edge of the crater."

   He could not visit the rice fields around Vercelli without mentioning that it was the native land of Virgil, who devoted his classical poems to agriculture. "The *Georgics* of Virgil is not only a poem but a significant document of historical importance, unsurpassed with respect to the amount of information it contains and the capacity of the author for observation. Just as 2000 years ago, the farmers listened to the verses of this agricultural poem, so today the *Georgics* can still serve as a manual for any farmers in Italy."
6. N. I. Vavilov to G. Karpechenko, February 23, 1926, *Nauchnoe Nasledstvo*, vol. 5, no. 375.

CHAPTER 13: THE BAREFOOT SCIENTIST

1. Lysenko's background is taken from several sources. They include Soyfer, *Lysenko and The Tragedy of Soviet Science*, Medvedev, *Rise and Fall*, and Reznik, *Nikolai Vavilov*.
2. Roll-Hansen, *Lysenko Effect*, p. 54.
3. Including Professor Nikolai Tulaikov (who had worked with Nikolai Ivanovich at Saratov) and was then director of the South-East USSR Institute of Cereal Culture.

4. See Soyfer, *Lysenko and the Tragedy of Soviet Science*, pp. 22–23. Gassner soaked the seeds in moist filter papers in Petri dishes in his laboratory and then put them in a refrigerator for about three weeks, during which time they started to germinate. This incipient germination altered the way the plant develops, reducing the number of leaves that have to unfold before the plant can flower from about twenty-two to twelve. Normally, in a winter strain of wheat, seeds planted in the autumn germinate before the onset of winter, pass through a period of cold, and they flower in the spring after producing only about twelve leaves and, thus, can produce grain before the summer is over.

5. Roll-Hansen, *Lysenko Effect*, p. 65, and Joravsky, *Lysenko Affair*, p. 60.

6. Soyfer, *Lysenko and the Tragedy of Soviet Science*, p. 13.

7. Especially by the young generation of radical British natural scientists, including John Desmond Bernal, J. B. S. Haldane, and Joseph Needham.

8. R. A. Gregory, *Discovery, Or the Spirit and Service of Science* (London: Macmillan, 1916), p. 41; and on Mendel, pp. 193–95.

9. Reznik, *Nikolai Vavilov*, pp. 262–67. Also in Roll-Hansen, *Lysenko Effect*, p. 58.

10. Reznik, *Nikolai Vavilov*, pp. 262–67.

11. Medvedev, *Rise and Fall*, p. 12.

12. Popovsky, *Vavilov Affair*, p. 52.

13. D.V. Ter-Avenesyan, in N. I. Vavilov, *Ocherki*, p. 206.

14. N. I. Vavilov to N. M. Tulaikov, *Nauchnoe Nasledstvo*, vol. 5, no. 98.

15. A. N. Ipatiev, in N. I. Vavilov, *Ocherki*, , pp. 91–107.

16. N. I. Vavilov to N. Gorbunov, *Nauchnoe Nasledstvo*, November 24, 1927, vol. 5, no. 475.

CHAPTER 14: THE GREAT BREAK

1. Joravsky, *Lysenko Affair*, p. 60.

2. Medvedev, *Rise and Fall*, p. 14.

3. Soyfer, *Lysenko and the Tragedy of Soviet Science*, p. 19.

4. The part played by Lysenko in these experiments relies on the two *Pravda* newspaper reports. The first, on July 21, 1929, had Lysenko preparing two sacks of seed for his father and then contacting the Ukraine officials and telling them about the bumper crop. The second, ten weeks later on October 8, 1929, has Lysenko stopping at his father's farm on the way to the conference, his father vernalizing the seeds and then telling the officials

about his success. In his own account, published six years later, Lysenko wrote, "This sowing was not accidental. It was conducted at my suggestion, by my father, D. N. Lysenko, at his farm." See Soyfer, *Lysenko and the Tragedy of Soviet Science*, pp. 18–19, plus notes pp. 311–12.

5. Roll-Hansen, *Lysenko Effect*, p. 90.

6. Vavilov, *Five Continents*, p. 72.

7. Birstein, *Perversion of Knowledge*, p. 43.

8. Ibid, p. 44.

9. Medvedev, *Soviet Agriculture*, p. 73.

10. Joravsky, *Lysenko Affair*, p. 37.

11. Ibid., p. 61.

12. N. I. Vavilov to G. Karpechenko, January 7, 1930, *Nauchnoe Nasledstvo*, vol. 10, no. 79, p. 65.

CHAPTER 15: STATE SECURITY FILE 006854

1. The file opened by the OGPU was No. 006854. See Rokityansky et al., *Sud Palacha*, p. 125. This file would eventually contain seven volumes. It would be summarized as an Investigation File in 1939 by Captain Zakharov, head of 3rd Economic Division of Ukrainian NKVD, and Lieutenant of State Security Makeev, in the NKVD (successor to the OGPU) memorandum on agents investigation, no. 139-1939. Central Archive of FSB of Russia, # R-2311, vol. 8, L. 24–26; also in Rokityansky, *Sud Palacha*, pp. 228–51. A summary of the surveillance report on Nikolai Vavilov is in a document entitled "Case No. P-1468, Memorandum on Agents Investigation N. 139-1929 'Selectionists.'" This memo was signed in August 1940 by Captain Zakharov, head of the Ukrainian NKVD, and Lieutenant Makeev. It appears to be a memo supporting the arrest document of Nikolai Vavilov, and mentions the first entry in Vavilov's file on March 11, 1930.

2. See Birstein, *The Perversion of Knowledge*, pp. 24–28. This pattern of gathering "evidence" was clear by the beginning of 1919, when a special department of the Cheka was created to weed out counterrevolutionaries in the middle class, the intelligentsia, and the clergy. A year later, the Cheka opened the first political show trial after its agents claimed to have unearthed the "Tactical Center," an underground group that had allegedly plotted to overthrow the Soviet state, and had contacts with a British spy, Paul Dukes, a former assistant conductor of the Maryinsky Theater in Petrograd. Dukes did work for British intelligence in Russia. One thou-

sand "members" of the group were arrested and sixty-seven executed. More would have been shot had not Trotsky and Lenin stepped in and commuted the sentences to imprisonment. Some of the defendants were sent to *sharashki*, special institutions for professionals, especially scientists, where they could continue their work, but in a concentration camp.

Then Lenin exiled or deported a long list of the intelligentsia, this time writers and professors, who were told they would be shot if they returned. The Cheka drew up the list of names from Moscow, Petrograd, and Ukraine approved by the Politburo and the lists were published in *Pravda*.

3. Ibid. Allegedly, the TKP was led by the agricultural economist Professor Nikolai Kondratiev. He had been assistant minister of food supply in the Provisional Government and was then director of the Market Institute within the USSR Commissariat of Finance. According to the newspaper reports he had been assisted by another official from the finance commissariat and Alexander Chayanov, who had been director of the Institute of Agricultural Economy, and also by Alexei Doyarenko, a professor at the Petrovka.

4. Popovsky, *Vavilov Affair*, p. 137.

5. Ibid., p. 138.

6. N. I. Vavilov, *Nauchnoe Nasledstvo*, Vol. 10, 1929–40, p. 77.

7. Ibid.

8. Medvedev, *Rise and Fall*, p. 18; also in Soyfer, *Lysenko and the Tragedy of Soviet Science*, p. 114.

9. Joravsky, *Lysenko Affair*, p. 77.

10. Ibid.

11. Captain Zakharov, Memorandum of Agents Investigation, 1939.

CHAPTER 16: THE PASSIONATE PATRIOT

1. N. I. Vavilov, *Five Continents*, p. 147.

2. Joravsky, *Lysenko Affair*, p. 26.

3. N. I. Vavilov to G. Karpechenko, January 7, 1930, *Nauchnoe Nasledstvo*, vol. 10, p. 65.

4. Ibid.

5. N. I. Vavilov to G. Karpechenko, May 10, 1930, *Nauchnoe Nasledstvo*, vol. 10, pp. 75–77.

6. Theodosius Dobzhansky, "N. I. Vavilov: A Martyr of Genetics,

1887–1942," *Journal of Heredity* 38 (August 1947), 227–32, quoted in Soyfer, *Lysenko and the Tragedy of Soviet Science*, p. 52. Some have suggested that Vavilov could only have been writing about the Soviet Union in such glowing terms to impress the OGPU agents who were opening his mail. However, there is evidence that this is not the case. Vavilov began writing even before he left America to Dobzhansky urging him to come back because his life would be better there and he could be more productive.

7. These letters appear in Vol. 3 of Nikolai Vavilov's *International Correspondence*, 1931–33, from January 1, 1931, to June 11, 1931, pp. 12–33.

8. T. G. Dobzhansky to N. I. Vavilov, *International Correspondence*, Vol. 3, p. 173.

9. Homer Shantz, quoted in Cohen, "Nikolai Ivanovich Vavilov," pp. 101–2.

10. Shantz letter quoted in ibid.

CHAPTER 17: A MODEST COMPROMISE

1. N. I. Vavilov to Y. Yakovlev, *Nauchnoe Nasledstvo*, vol. 10, p. 163 and 172.

2. Roll-Hansen, *Lysenko Effect*, letter to Lenin Academy presidium, p. 92.

3. Krementsov, *Stalinist Science*, p. 27.

5. Bukharin had been editor of *Pravda* for the last five years.

6. Bukharin also supported Mendelian genetics as a basis for plant breeding. It was an endorsement that would stand as another mark against Vavilov in the hard years ahead.

7. N. I. Vavilov to Y. I. Barulina, June 23, 1931, Yuri Vavilov private collection.

8. Nikolai Bukharin, in Science at the Crossroads: Papers presented to the International Congress of the History of Science and Technology, London, June 29 to July 3, 1931, p. 16.

9. Nikolai Vavilov, in Science at the Crossroads: Papers presented to the International Congress of the History of Science and Technology, London, June 29 to July 3, 1931, p. 106.

10. Ibid, p. 97.

11. Quoted in N. Roll-Hansen, *Lysenko Effect*, p. 78; from the journal, *Socialist Reconstruction of Agriculture*, joint publication of the Lenin Academy of Agriculture and the Commissariat of Agriculture, 56, pp. 72–73.

12. See Joravsky, *Lysenko Affair*, pp. 86–87; and Soyfer, *Lysenko and the Tragedy of Soviet Science*, p. 34.

13. Joravsky, *Lysenko Affair*, p. 87.

14. Soyfer, *Lysenko and the Tragedy of Soviet Science*, p. 28. Such urgency gave more weight, at least publicly, to Yakovlev's refusal to allow Vavilov to travel. His planned trip to India was now canceled and Nikolai Ivanovich, who had always assumed that India was the center of origin of rice, would not be given a chance to demonstrate his case. For those at the Institute, and even for Lenochka, he appeared to take this setback in his stride, but for the world's premier plant hunter, the loss of this opportunity was particularly devastating.

15. Roll-Hansen, *Lysenko Effect*, p. 133.

CHAPTER 18: THE RED PROFESSOR

1. Berg, *Acquired Traits*, pp. 29–30. One of Prezent's students was the young geneticist Raissa Berg, whose father, Lev Berg, was a geography professor and a friend of Vavilov's. Prezent had become a confirmed Lamarckist, like Stalin. In his lectures Prezent argued that external factors played a greater part in heredity than genes, and when Raissa Berg challenged him, referring to recent genetic research, he replied, "I don't read specialized literature, I only criticize science's fundamental orientations." Instead of engaging in a scientific debate, Prezent called on Berg to "tell us about the Marxist-Leninist theory of cognition."

"I can't do it without preparation," she protested.

"Do you refuse to answer?" demanded Prezent.

"Yes," she said.

Prezent tried to have Berg expelled for her lack of Marxist knowledge, and when that failed he had the incident written up in the university newspaper, just to embarrass her.

2. Medvedev, *Rise and Fall*, p. 10.

3. Soyfer, *Lysenko and the Tragedy of Soviet Science*, p. 62, suggests that Prezent worked at Vavilov's Institute in Leningrad at the beginning of Stalin's cultural revolution, and even his fellow ideologues found him difficult. But Tatyana Lassan, for many years the Institute's archivist, found no evidence of this, and Nikolai Ivanovich himself never mentioned Prezent's brief stay—he may have been away on one of his expeditions.

4. T. D. Lysenko, Review of scientific work of Prezent, I. I., memorandum

to Lenin Academy, Archive of Academy of Sciences of USSR, approx. 1936, Fond 1593, opis 1. This archive, which was microfilmed in 1999 by Research Publications, an imprint of Primary Source Media, of Woodbridge, Connecticut, is open at the British Library in London under the title History of Russia's Study of Genetics. The documents are now held by the Archive of the Russian Academy of Sciences, Moscow.

5. T. D. Lysenko, Review of scientific work of Prezent. See previous note. Soyfer, *Lysenko and the Tragedy of Soviet Science*, p. 63.

6. Joravsky, *Lysenko Affair*, p. 237.

7. Ibid., p. 238.

8. T. D. Lysenko, Review of Scientific work of Prezent.

9. N. I. Vavilov to T. Lysenko, *Nauchnoe Nasledstvo*, vol. 10, 1929–40, March 29, 1932, p. 165.

10. N. I. Vavilov to N. V. Kovalyov, August 9, 1932, *Nauchnoe Nasledstvo*, vol. 10, p. 179.

11. Roll-Hansen, *Lysenko Effect*, p. 160.

12. N. I. Vavilov to A. N. Kovalyov, *Nauchnoe Nasledstvo*, vol. 10, 1929–40, August 9, 1932, p. 179.

CHAPTER 19: THE LAST EXPEDITION

1. Krementsov, *Stalinist Science*, p. 34.

2. N. I. Vavilov to Ye. I. Barulina, August 15, 1932, private collection of Yuri Vavilov.

3. Ibid.

4. Proceedings of the Sixth International Congress of Genetics, Ithaca, New York, 1932, vol. 1 (New York: Brooklyn Botanic Garden), p. 332.

5. Ibid., p. 333.

6. Ibid., p. 336.

7. Ibid., p. 340.

8. Ibid., p. 342.

9. N. I. Vavilov to N. V. Kovalev, *Nauchnoe Nasledstvo*, vol. 10, November 7, 1932, p. 185, quoted in Loskutov, *Vavilov and His Institute*, p. 76.

10. Ibid.

11. N. I. Vavilov, *Five Continents*, p. 156. He even picked out potatoes that would still be edible after a period of freezing—so that they could be preserved "for a few years."

12. I. Loskutov, *Vavilov and His Institute*, p. 76.

13. N. I. Vavilov, *Five Continents*, p. 157.

14. Ibid., p. 134.

15. Ibid., p. 138.

16. Ibid., p. 144.

17. Ibid., p. 141.

18. There were other factors bearing on his outlook. Vavilov was traveling in Europe and America during the Great Depression, going out of the cities into the countryside and witnessing the effects of the economic disaster firsthand. (There were fourteen million unemployed in America.) Vavilov received several applications from socialist biologists in Europe and America seeking work in the Soviet Union.

19. Captain Zakharov, NKVD Memorandum, "Links with Abroad," 1939.

20. *Paris-Midi*, February 14, 1933.

21. Captain Zakharov, NKVD Memorandum, "Links with Abroad," 1939.

22. Elof Axel Carlson, *Genes, Radiation and Society: The Life and Work of H. J. Muller* (Ithaca: Cornell University Press, 1981), p. 189.

23. Ibid., p. 192.

CHAPTER 20: THUNDER AND DRAGONS

1. Popovsky, *Vavilov Affair*, p. 72.

2. N. I. Vavilov to N. M. Tulaikov, April 28, 1933, *Nauchnoe Nasledstvo*, vol. 10, p. 195.

3. E. S. Yakushevsky, "Brothers Nikolai and Sergei Vavilov, The Past and Thoughts," Memorial Evening, January 6, 1989, St. Petersburg Division of the Russian Culture Foundation, 1994.

4. N. I. Vavilov to A. Sapegin, April 6, 1933, *Nauchnoe Nasledstvo*, vol. 10.

5. E. N. Sinskaya, *Vospominanya o N. I. Vavilova [Recollections of N. I. Vavilov]* (Kiev: Naukova Dumka, 1991), p. 152.

6. Ibid.

7. Ibid., p. 153.

8. Birstein, *Perversion of Knowledge*, p. 213.

9. Captain Zakharov, NKVD Memorandum, "Hostility to the Soviet Regime," 1939.

10. N. I. Vavilov to L. I. Govorov, July 31, 1931, *Nauchnoe Nasledtsvo*, vol. 10, p. 126.

11. FSB Russia Central Archive, No. P-2311, vol. 8, L 33-40. In Y. Rokityansky et al., eds., *Sud Palacha* (Moscow, 1999), pp. 158–63.

12. Medvedev, *Rise and Fall*, p. 30, for Russian experiments in mutagenesis.

13. Carlson, *Genes, Radiation and Society*, p. 193.

14. Ibid., p. 194.

15. Ibid., p. 178.

16. Berg, *Acquired Traits*, p. 28.

17. N. I. Vavilov to Party Committee of the Sector of Selection and North-West Center of Selection in Detkse Selo, January 31, 1934, in *Nauchnoe Nasledstvo*, vol. 10, p. 217.

18. It is generally thought that Yakovlev was behind the cancellation. At the time, Yakovlev was moving from the Ministry of Agriculture to a more powerful position as head of the Agriculture Department of the Central Committee. As Yakovlev's support of Lysenko was increasing, his relationship with Vavilov had been deteriorating.

19. N. I. Vavilov to Ia. Yakovlev, March 23, 1932, *Nauchnoe Nasledstvo*, vol. 10, pp. 163–64.

20. N. I. Vavilov to T. D. Lysenko, May 26, 1934, *Nauchnoe Nasledstsvo*, no. 381, p. 233.

21. V. B. Yenkin in N. I. Vavilov, *Ocherki*, p. 337.

22. E. Sinskaya, *Recollections of N. I. Vavilov*, pp. 155–56.

CHAPTER 21: THE LYSENKO OFFENSIVE

1. At the 2nd All-Union Congress of Collective Shock Workers, February 14, 1935, in Roll-Hansen, *Lysenko Effect*, p. 97.

2. Ibid.

3. Cohen, "Nikolai Ivanovich Vavilov," p. 195, quoting an article, "Death of a Science in Russia," *The Listener*, London, December 9, 1948, p. 873.

4. Ibid.

5. Ibid., p. 58.

6. Soyfer, *Lysenko and the Tragedy of Soviet Science*, p. 35.

7. Roll-Hansen, *Lysenko Effect*, p. 97.

8. In the rising tensions between the agricultural bureaucracy and the political leadership, the younger generation sided with the political leadership.

9. Bondarenko and Klimov to Stalin, March, 27, 1935, document #3 in Rokityansky et al., *Sud Palacha*, p. 164.

10. Roll-Hansen, *Lysenko Effect*, pp. 97–99.

11. Zhores Medvedev and Roy Medvedev, *The Unknown Stalin* (London: I. B. Taurus, 2003), p. 183.

12. See Cohen, "Nikolai Ivanovich Vavilov," p. 203, and Medvedev, *Rise and Fall*, p. 15.

13. Roll-Hansen, *Lysenko Effect*, p. 173. Vavilov did criticize Lysenko's idea that the early-ripening gene was always dominant over late; see Roll-Hansen, *The Lysenko Effect*, p. 170.

14. Lysenko's idea presented a major problem for naturally self-pollinating plants, like wheat, where the female egg is fertilized by pollen from the same plant. Lysenko's absurd solution was for workers to snip off the flowers with a pair of scissors, extract the pollen, and carry out the crossbreeding with the aid of a paintbrush. Prezent even invented a bumper-sticker term for this procedure: a "marriage of love"—as opposed to the natural self-pollination, which was a "forced marriage." Vavilov opposed Lysenko, pointing out that such methods would destroy established varieties. Other scientists argued that a seed produced in such a way could not be called a "pure" or "elite" seed because it would contain a new, and unknown, mixture of genes. Even so, the criticism in no way matched the increasing ferocity of Lysenko's attacks designed to denigrate academic scientists and praise the peasant farmers who practiced "kolhoz [collective farm] and sovkhoz [state farm]" horticulture.

15. Soyfer, *Lysenko and the Tragedy of Soviet Science*, p. 69.

16. Ibid., p. 70.

17. Ibid., pp. 72–73.

CHAPTER 22: THE SHOWDOWN

1. Agol worked at the Medical Genetics Institute in Moscow. His colleague Solomon Levit, the director of the institute, had also worked with Muller in Texas. Levit also was a party member and had been arrested in November. He would also be shot.

2. Soyfer, *Lysenko and the Tragedy of Soviet Science*, p. 88, quoting A. Prokofoyeva at a meeting of USSR society of Geneticists and Breeders.

3. Roll-Hansen, *Lysenko Effect*, p. 196.

4. Soyfer, *Lysenko and the Tragedy of Soviet Science*, p. 83.

5. There was a need, another speaker said, for his fellow geneticists to be even tougher on Lysenko—to be more forceful in their analysis of his work and to expose his mistakes. This should be done in public, not just in private conversation, as had been done in the past. When a geneticist sees a "jumble of nonsense in his opponent's speeches, he can no longer afford to be indulgent." Soyfer, *Lysenko and the Tragedy of Soviet Science*, p. 84.

6. Roll-Hansen, *Lysenko Effect*, p. 197.

7. Medvedev, *Rise and Fall*, p. 35.

8. Soyfer, *Lysenko and the Tragedy of Soviet Science*, p. 87.

9. Medvedev, *Rise and Fall*, p. 28.

10. Muller's lecture was read by Nikolai Koltsov, a zoologist, known as the grand old man of Russian experimental biology.

11. Carlson, *Genes, Radiation and Society*, p. 232.

12. H. Muller to M. Popovsky, personal communication, June 16, 1966, p. 2. Copy in Y. Vavilov's private collection.

13. Ibid.

14. Ibid.

15. Ibid.

16. Igor Loskutov, personal communication from Otto Frankel, in *Vavilov and His Institute*, p. 50.

17. J. Stalin to V. Molotov on Central Committee Resolution of August 2, 1935, on 7th International Congress of Genetics in the USSR, in V. D. Esakov, Academy of Sciences in decisions of Politburo CC RKP (b)-VKP (b)-KPSS, 1922–52 (Moscow: Rosspen, 2000).

18. K. Bauman to J. Stalin and V. Molotov on Central Committee Resolution of August 2, 1935 on the 7th International Congress of Genetics in the USSR, in V. D. Esakov, Academy of Sciences in decisions of the Politburo CC RKP (b)-VKP (b)-KPSS, 1922–52 (Moscow: Rosspen, 2000).

19. *New York Times*, December 17, 1936, p. 26.

20. N. I. Vavilov, cable to *New York Times*, December 22, 1936; appeared in full in *Izvestia*, December 22, 1936, p. 4.

CHAPTER 23: THE TERROR

1. Maria Zaitseva on a visit to N. I. Vavilov in his Moscow apartment, in *Brothers Nikolai and Sergei Vavilov* (Moscow: Lebedev Institute, 1994), pp. 40–45. Also, Soyfer, *Lysenko and the Tragedy of Soviet Science*, p. 130.

2. M. G. Zaitseva, memorial speech at St. Petersburg branch of Russian Culture Fund, January 6, 1989, in *Brothers Nikolai and Sergei Vavilov*, pp. 40–45.

3. L. Kursonova, in Y. Rokityansky, "Genius's Son," *Chelovek*, vol. 6, 2003, p. 111.

4. Koltsov's institute had become the central research institution for the study of cells, and in 1928 Koltsov had proposed the idea that hereditary traits resided in special molecules—anticipating the double helix by twenty-five years. He was the leader of experimental biology in the Soviet Union,

and, as a geneticist, he was an obvious target for the Lysenko forces—even though he had supported Lysenko's vernalization, and had been a liberal fighting for students' rights under the tsar. He had been arrested during the first Soviet show trial against the so-called Tactical Center in 1920. In 1939, Koltsov was a candidate, with Lysenko, to become a full member of the Academy of Sciences. Koltsov, whose candidacy was supported by Vavilov, was defeated. Lysenko became a full member. Koltsov was not arrested. He died of a heart attack in 1940. His name was removed from scientific literature for twenty-five years. Soyfer, *Lysenko and the Tragedy of Soviet Science*, p. 124.

5. Ibid., p. 104; Medvedev, *Rise and Fall*, p. 49.
6. Medvedev, *Rise and Fall*, p. 46.
7. Ibid., p. 48.
8. Ibid., p. 53.
9. See Medvedev: "This was a conscious, organized, and purposeful baiting, in a premeditated way, to expose scientific opponents to the blows of the punitive organs of our country." Ibid., p. 50.
10. Georgy Meister, a plant breeder and critic of Lysenko, went mad in jail. See Birstein, *Perversion of Knowledge*, and Soyfer, *Lysenko and the Tragedy of Soviet Science*, for more on the arrests.
11. See Jack Harlan, *The Living Fields* (Cambridge, England: Cambridge University Press, 1995).
12. Cohen, "Nikolai Ivanovich Vavilov," p. 220. Also in "Jack Harlan, Plant Explorer," www.biodiversityinternational.org/publications
13. H. J. Muller to J. S. Huxley, March 13, 1937, Lilly Library.
14. N. I. Vavilov to H. J. Muller, May 8, 1937, Lilly Library; quoted in Carlson, *Genes, Radiation and Society*, p. 239.
15. Ibid., p. 239.
16. H. J. Muller to J. S. Huxley, March 13, 1937, Lilly Library, quoted in ibid., p. 242.
17. H. Muller to M. Popovsky, personal communication, June 16, 1966, p. 3.
18. Berg, *Acquired Traits*, p. 40.
19. H. Muller to M. Popovsky, personal communication, June 16, 1966.

CHAPTER 24: INTO THE PYRE

1. Sergei Vavilov, Diaries, 1939–51, Voprosi Istorii Estestvoznaniya I Teckhniki (VIET), 2004, no. 2, pp. 3–52, entry April 2, 1941, p. 10.

2. G. Shlykov to Central Committee of USSR, science section, 1938, quoted in Soyfer, *Lysenko and the Tragedy of Soviet Science*, p. 119.
3. See Soyfer, p. 119, and *Ogonyok*, no. 47, Moscow, Nov. 1987, p. 6.
4. Soyfer, *Lysenko and the Tragedy of Soviet Science*, p. 123.
5. J. V. Stalin, Speech in the Kremlin for Workers in Higher Education, May 17, 1938, in Soyfer, *Lysenko and the Tragedy of Soviet Science*, p. 122.
6. Quoted in Popovsky, *Vavilov Affair*, p. 152.
7. Roll-Hansen, *Lysenko Effect*, p. 252.
8. N. I. Vavilov to G. Karpechenko, October 10, 1938, *Nauchnoe Nadledstvo*, vol. 10, no. 694, p. 384.
9. Soyfer, *Lysenko and the Tragedy of Soviet Science*, p. 127.
10. Ibid., p. 128.
11. Ibid.
12. N. I. Vavilov to H. Muller, March 7, 1938, *International Correspondence*, vol. 6, p. 60.
13. N. I. Vavilov to S. C. Harland, London, September 13, 1938, *International Correspondence*, vol. 6, p. 94.
14. N. I. Vavilov to H. J. Muller, December 18, 1938, *International Correspondence*, vol. 6, p. 107.
15. N. I. Vavilov to F. E. Clements, Washington, D. C., March 22, 1938, *International Correspondence*, vol. 6, p. 67.
16. Ibid.
17. Sinskaya in Popovsky, *Vavilov Affair*, p. 93.
18. J. Hawkes, personal communication with I. Loskutov, in Loskutov, *Vavilov and His Institute*, pp. 51–53.
19. In Medvedev, *Rise and Fall*, pp. 57–58.

CHAPTER 25: COMRADE PHILOSOPHERS

1. Medvedev, *Rise and Fall*, pp. 59–63.
2. Popovsky, *Vavilov Affair*, p. 119.
3. E. S. Levina, Vavilov et al. *Biology in USSR: History and Historiography*, (Moscow, AIRO, 1995).
4. Organized by the editorial board of *Under the Banner of Marxism* (theoretical journal of the Communist Party), October 7–14, 1939.
5. Soyfer, *Lysenko and the Tragedy of Soviet Science*, p. 133, and Medvedev, *Rise and Fall*, p. 64.
6. Soyfer, *Lysenko and the Tragedy of Soviet Science*, p. 134.

7. See N. Krementsov, *Stalinist Science*, p. 76, for commentary and references for the letter. Also, Y. Rokityansky et al., *Sud Palacha*, p. 51.

8. Rokityansky et al., *Sud Palacha*, p. 66.

9. N. I. Vavilov letter to K. Pangalo, *Nauchnoe Nasledstvo*, vol. 10, 1939, p. 417.

10. N. I. Vavilov to H. J. Muller, June 12, 1939, *International Correspondence*, p. 110.

11. Soyfer, *Lysenko and the Tragedy of Soviet Science*, p. 135. Soyfer says that he was given this account of the meeting with Stalin by Yefrem S. Yakushevsky, who worked at Vavilov's Institute for many years and who spoke to Vavilov on November 28, 1939, eight days after the meeting with Stalin. The recollection was recorded by D. V. Lebedev, who told it to Soyfer in 1987. Zhores Medvedev, who with his brother, the Russian historian and Stalin specialist Roy Medvedev, has examined Stalin's appointments book, told me there is no record of this meeting—and the record of Stalin's visitors was uncommonly thorough. Whether the meeting took place is therefore in doubt.

12. Cohen, "Nikolai Ivanovich Vavilov," p. 225.

13. N. I. Vavilov to Ivan Benediktov and secretary of the Central Committee, A. Andreyev, undated 1940, *Ogonyok*, November 1987, Letter no. 2, Moscow.

14. Ibid.

15. I. K. Fortunatov, in N. I. Vavilov, *Ocherki*, p. 367.

16. S. I. Vavilov diaries, April 3, 1940.

CHAPTER 26: THE ARREST

1. Yelena had developed what is known as Erb's palsy, a condition that can affect one or all of the five primary nerves supplying movement and feeling to an arm. In adults it can be caused by a traumatic fall on the side of the neck.

2. V. Soyfer, *Lysenko and the Tragedy of Soviet Science*, p. 136, from a tape recording made by Soyfer of a talk by A. Prokofyeva-Belgovskaya at the N. I. Vavilov Society of Geneticists and Breeders, Moscow, 1983.

3. N. I. Vavilov to S. C. Harland, June 22, 1940. *International Correspondence*, vol. 6, p. 116.

4. Loskutov, *Vavilov and His Institute*, pp. 34–39.

5. Yuri Vavilov, private archive.

6. N. I. Vavilov to Doncho Kostov, May 25, 1940.

7. Popovsky, *Vavilov Affair*, p. 120.

8. Ibid., p. 112.

9. Soyfer, *Lysenko and the Tragedy of Soviet Science*, p. 139.

10. Popovsky, *Vavilov Affair*, p. 124.

11. Ibid., p. 125.

12. Lydia Breslavets, quoted in Cohen, "Nikolai Ivanovich Vavilov," p. 226, and Popovsky, *Vavilov Affair*, p. 125.

13. Popovsky, *Vavilov Affair*, p. 125, and Cohen, "Nikolai Ivanovich Vavilov," p. 226.

14. Popovsky, *Vavilov Affair*, p. 127.

15. N. I. Vavilov to Oleg, August 2, 1940.

16. Protocol of search of the apartment of citizen Vavilov N. I. at Ul. Chakalovskaya, no. 21, apt. no. 57, NKVD Document no. 17, August 7, 1940, Rokityansky, *Sud Palacha*, p. 216. The list of confiscated items also includes "a letter from Vavilov to Stalin on disagreement with Academician Comrade Lysenko, 10 pages, and a letter from Vavilov to VKP (b) Central Committee on scientific questions of selection, 21 pages." None of this material survived.

17. Popovsky, *Vavilov Affair*, p. 165.

18. Sergei Vavilov, Diaries, August 13, 1940.

19. Lidia Kurnosova, in Y. Rokityansky, "Genius's Son."

20. Ibid.

21. Sergei Vavilov, Diaries, August 14, 1940.

CHAPTER 27: THE INTERROGATION

1. Rokityansky et al., *Sud Palacha*, p. 462.

2. It is not known exactly when Stalin approved the arrest, but it is possible that he did so during a meeting with Beria in his Kremlin office. See "Istorichesky Arhiv," 1996, #2, p. 24. On August 6, Beria endorsed the arrest warrant. See Yuri Vavilov, *Long Search*, p. 136.

3. Yuri Vavilov, *Long Search*, p. 133.

4. Birstein, *Perversion of Knowledge*, p. 210.

5. Ibid., p. 219.

6. Rokityansky et al., *Sud Palacha*, and Yuri Vavilov, *Long Search*.

7. N. I. Vavilov in Protocol of Interrogation, Document no. 38, August 24, 1940, also in Y. Rokityansky et al., *Sud Palacha*, pp. 269–71.

8. Rokityansky et al., *Sud Palacha*, p. 13.

9. Ibid., p. 79.

10. Ibid., p. 80.

11. N. I. Vavilov, in Protocol of Interrogation, Document no. 39, August 27–28, 1940, Rokityansky et al., *Sud Palacha*, p. 271.

12. S. I. Vavilov, diaries, December 31, 1940.

13. Rokityansky et al., *Sud Palacha*, p. 89.

14. This interrogation session took place on June 25, 1941.

15. Rokityansky et al., *Sud Palacha*, p. 87.

16. Document 85 of the NKVD Protocol, interrogation of Propkopy Maximovich Lobov by Marisov, Lieutenant of State Security; Rokityansky et al., *Sud Palacha*, p. 24; 1941, Central Archive of Russian FSB, no. R-2311, vol. 5, 1, 185–86.

17. Rokityansky et al., *Sud Palacha*, p. 92.

18. Ibid., p. 93.

19. Birstein, *Perversion of Knowledge*, p. 229.

20. Military Collegium of the Supreme Court of the USSR, sentence of N. I. Vavilov, July 9, 1941, in Yuri Vavilov, *The Long Search*, p. 142.

21. N. I. Vavilov in a letter from Saratov jail to Beria, June 25, 1942. Vavilov asserted that he had "categorically denied" all charges against him.

22. Y. Rokityansky et al., *Sud Palacha*, p. 516.

CHAPTER 28: RETURN TO SARATOV

1. Popovsky, *Vavilov Affair*, p. 175, interview with one of the convicts, Dr. Andrei Ivanovich Sukhno.

2. Ibid., p. 182.

3. Ibid.

4. Yuri Vavilov, interview with the author, Moscow, July 2004.

5. Popovsky, *Vavilov Affair*, p. 183.

6. Sinskaya, in ibid., p. 159.

7. Irina Yankovskaya-Piotrovskaya in Yuri Vavilov, *Long Search*, p. 167.

8. N. I. Vavilov to L. Beria, April 25, 1942, Rokityansky et al., *Sud Palacha*, pp. 519–21; FSB Russia Central Archive, no. P-2311, v, 1, L495–461.

9. See letter from Sir Henry Dale to ANSSR, 1945, in *Ogonyok*, 47, Moscow, November 1987.

10. Yuri Vavilov archive.

11. There has been speculation over what kind of work Vavilov might have been asked to do if he had been sent to a labor camp. One possibility was development of the Russian dandelion for producing natural rubber. It was an idea being pondered in Hitler's Germany, where, like Russia, there was no natural substitute for this vital material.

12. Medvedev, *Rise and Fall*, p. 72.

13. S. Dyachenko, *Ogonyok*, no. 47, November 1987, pp. 1–2.

14. Sergei Vavilov, Diaries, August 29, 1941; March 15, 1942; April, 26, 1942; October 2, 1942.
15. Yuri Vavilov, interview with the author.
16. Medvedev, *Rise and Fall*, p. 74.
17. Document 129, Protocol 137 of the NKVD file, Rokityansky et al., *Sud Palacha*, p. 526.

CHAPTER 29: "OLEG, WHERE ARE YOU?"

1. Author interview with Yuri Vavilov, Moscow, 2004.
2. S. I. Vavilov, Diaries, July 5, 1943.
3. S. I. Vavilov, Diaries, July 6, 1943.
4. Ibid.
5. N. A. Tolstoy, *Brothers Nikolai and Sergei Vavilov* (RAN, 1994), p. 8.
6. S. I. Vavilov, Diaries, October 26, 1943.
7. Author interview with Yuri Vavilov, Moscow, 2004.
8. Popovsky, *Vavilov Affair*, p. 195.
9. S. I. Vavilov, *Ocherki I Vospominaniya* (1991), p. 156.
10. S. M. Alexanyan and V. I. Krivchenko, [article], *Diversity*, April, 1992; see also *The Washington Post*, May 12, 1992, p.1.
11. S. I. Vavilov, Diaries, May 20, 1944.
12. Zh. and R. Medvedev, *Unknown Stalin*, p. 197.
13. S. I. Vavilov, 1891–1951, *Epokha i Lichnost', Fiziki, Ocherki i Vospominaniya* (Moscow: Nauka, 1999), pp. 137–75.
14. Yuri Vavilov, *Long Search*, p. 20.
15. Rokityansky, "Genius's Son," pp. 141–59.
16. I. A. Zaharov-Gezehus to R. G. Nurgaliev, Russian Federation Minister of Interior, April 5, 2007. (Source: Russian Academy of Science, Committee on Preservation and Development of Scientific Heritage of N. I. Vavilov.)
17. Ibid.

EPILOGUE: VAVILOV'S GHOST

1. See Medvedev, *Soviet Agriculture*, p. 132, and Krementsov, *Stalinist Science*, p. 160.
2. Quoted in Medvedev, *Soviet Agriculture*, p. 134, from N. S. Khrushchev, *Khrushchev Remembers* (London: Andre Deutsch, 1971), pp. 200–215.
3. Medvedev, *Soviet Agriculture*, p. 132.

4. Roll-Hansen, *Lysenko Effect*, p. 268.

5. The main official attack came from Yuri Zhdanov, who was in charge of the science department of the Central Committee. A young chemist, he was the son of Politburo member Andrei Zhdanov. See Medvedev, *Unknown Stalin*, pp. 182–84.

6. Roll-Hansen, *Lysenko Effect*, p. 267, quoting Leslie Dunn, "Science in the USSR: Soviet Biology," *Science* 99 (January 28, 1944), pp. 65–67.

7. M. Flintner, *Sammler, Rauber und Gelehrte: Die Politischen Interessen an Pflanzengenetischen Resourcen, 1895–1995* (Frankfurt: Campus Verlag, 1995), quoted in R. Pistorius and J. van Wijk, *The Exploitation of Plant Genetic Information* (New York and Wallingford, UK: CABI, 1999), p. 62. See also Carl-Gustav Thomstrom and Uwe Hossfeld, "Instant Appropriation—Heinz Bucher and the SS Botanical Collection Commando to Russia 1943," *Plant Genetic Resources Newsletter* 129 (2002), p. 39.

8. S. C. Harland and C. D. Darlington, "Obituaries, Prof. N. I. Vavilov, For. Mem. R.S.," *Nature* 3969 (November 24, 1945), pp. 621–22.

9. Quoted in Medvedev, *Unknown Stalin*, p. 186, from *Pravda*, February 10, 1946.

10. Eric Ashby, "Genetics in the USSR," *Nature* 158 (August 31, 1946), pp. 285–87.

11. Krementsov, *Stalinist Science*, p. 159.

12. See Medvedev, *Unknown Stalin*, p. 192.

13. Ibid., p. 193. Stalin's letter is dated October 31, 1947. It was found in the Kremlin Presidential Archive by Yuri Vavilov when he was searching for evidence of Stalin's personal involvement in his father's arrest.

14. Ibid.

15. Ibid., p. 186.

16. S. I. Vavilov, Diaries, July 15, 1948.

17. Stephen Jay Gould, *Hen's Teeth and Horses' Toes* (New York: W. W. Norton, 1994), p. 135, also quoted in Krementsov, *Stalinist Science*, p. 340 n. 59.

18. Efroimson was freed in 1955 and "rehabilitated" in 1956.

19. Soyfer, *Lysenko and the Tragedy of Soviet Science*, p. 198.

20. Sergei Vavilov, Diaries, July 15, 1949, p. 33.

21. Yuri Vavilov, interviewed by N. Alexandrova for the author, July 2007.

22. Yuri Vavilov, "Different Fates of Vavilov Brothers," *Chelovek*, 2002, no. 4, p. 144.

23. See Medvedev, *Rise and Fall*, p. 136.

24. Trofim Lysenko, letter to Office of the Chief Military Prosecutor, June 25, 1955; Yuri Vavilov Collection.

25. V. Semichastny to N. S. Khrushchev, classified memorandum, Archives of the Soviet Communist Party and Soviet State, Hoover Institution Archives, Microfilm Collection, reel no. 1, 1009, 7 opis 65 (12). In addition: In a classified letter from Yuri Andropov, chairman of the Committee of State Security (KGB), to the Central Committee, CPSU, dated December 8, 1976, Andropov informed the Central Committee that materials from Lysenko's file, including his correspondence with the CPSU CC, the CPSU Council of Ministers, and the USSR Academy of Sciences, 426 pages, not secret, had been removed with "the purpose of preventing them from getting to the West and used in anti-Soviet propaganda." The Federal Security Service (FSB) of the Russian Federation, suggested that these documents might have been transferred from the KGB archive to the Russian Government Archive of Contemporary History, RGANI. Efforts to obtain these files were unsuccessful. The RGANI reported that they were still classified, and, later, that the files were not in the RGANI archive.

26. Loren Graham, *Moscow Stories: The Biggest Fraud in Biology* (Bloomington: Indiana University Press, 2006), pp. 121–24.

27. H. Harlan to B. M. Cohen, personal communication, August 4, 1975, p. 1. Harlan was commenting on events in Leningrad at the Twelfth International Botanical Congress, July 9, 1975.

28. *Ogonyok*, Moscow, November 1987.

29. Yevgeniya Albats, *Genius and the Villains, Moscow News* weekly, no. 46, November 15, 1987, p. 10.

30. Letter from R. A. Noble to S. J. Butt, Soviet Department, Foreign Office, London, November 30, 1987.

31. *Dictionary of Scientific Biography*, vol. 15, Supplement I, Topical Essays (New York: Charles Scribner's Sons, 1978), pp. 505–13.

32. Jack Harlan, "Agricultural Origins: Centers and Non-Centers," *Science*, vol. 174 (1971), pp. 468–74; see also Harlan, *The Living Fields*.

33. See John Seabrook, "Sowing for Apocalypse: Annals of Agriculture," *The New Yorker*, August 13, 2007.

Acknowledgments

I was extremely fortunate in finding many people who have been touched by this tragic story and who went out of their way to help in its telling. The book could not have been written without the meticulous research and superb translations of Natalia Alexandrova. Her legendary doggedness in the face of government officials made all the difference at crucial moments. As the book went to press, Natalia was still trying to get archive documents declassified.

Alice Mayhew provided her special brand of enthusiasm, and uncommon wisdom. It was quite simply a privilege to write another book for her. Roger Labrie's fine eye prevented many a misstep, and Loretta Denner is a wonderful production editor. Michael Carlisle was passionate about the project from the beginning. I thank them all.

The Alfred P. Sloan Foundation generously awarded me a grant for travel and research in Russia from their Public Understanding of Science and Technology Program.

In Russia, Yuri Vavilov, Nikolai's son and the keeper of his father's archive, kindly gave me unpublished letters and photographs, and accompanied me on a tour of the places where his father lived and worked in St. Petersburg and nearby Pushkin.

Many Russians who have worked, and are still working, on the life of Nikolai Vavilov, directly or indirectly, were generous with their time and their sources. In Moscow, I thank Elena Levina and Vladimir Esakov, Ilya Zaharov-Gezehus, Yakov Rokityansky, Natalia

Dubrovina, Boris Altshuler, Denis Shibaev, Kirill Anderson, Irina Borovskih, and Elena Tyurina. In Ivashkovo, I was most fortunate to find Tamara Katanova, great-granddaughter of Ivan, brother of Ilya Vavilov, Nikolai's grandfather. Svetlana Radchenko took me on a tour of the village.

In St. Petersburg, I thank the officials and staff of the N. I. Vavilov Institute of Plant Industry, especially the former director, Viktor Dragavtsev, and also Professor Igor Loskutov, Sergei Alexanyan, Zoya Mikhailova, and Sergei Shuvalov. I am grateful to Professor Loskutov for his continued support and permission to use VIR photographs published in his book *Vavilov and His Institute* (1999). Tanya Lassan, the former VIR archivist, was very generous with her time and provided me with several hard-to-find Vavilov texts.

At St. Petersburg archives, Nadezhda Fomichyova and Professor Irina Tunkina were most welcoming and helpful.

In Russia, many friends gave me hospitality. I thank especially Boris and Masha Ryzhak in Moscow and Tamara Naumova in St. Petersburg. Galina Alexandrova kindly hosted a lunch for Yuri Vavilov.

In Europe, Carl-Gustaf Thornstrom kept me up to date with developments in the saga of the German SS Commando that stole some of Vavilov's precious seeds.

In London, my longtime friend Zhores Medvedev shared his memories and special insights of the Stalin period and Lysenkoism, and generously gave me Vavilov volumes from his Russian library. As always, he provided invaluable advice on Soviet politics, science, and agriculture. I thank Gina Douglas of the Linnean Society for giving me access to the Society's files on Vavilov. My thanks to the John Innes Horticultural Institute for help with the William Bateson papers, and to the staffs of the British Library, the Royal Botanic Library at Kew, and the Public Records Office at Kew.

In the United States, David Hoffman was extraordinarily generous in directing me to key documents at the Hoover Institution

Archives at Stanford, California, where he was researching his own book on the Cold War. Lora Soroka helped find the documents.

Valery Soyfer generously shared his interpretation and insights of Vavilov's life and work, and allowed me to publish photographs from his comprehensive work *Lysenko and the Tragedy of Soviet Science*. Mark Popovsky graciously offered to discuss his 1984 groundbreaking biography with its unique sources, but sadly we had not arranged a meeting before he died. Barry Mendel Cohen guided me to his important 1980 dissertation on Vavilov, and generously provided me with copies of other published material.

At the USDA's National Agricultural Library, I thank Maria Pisa and Susan Fugate. James Thompson provided documents on the 1939 International Congress of Genetics. The New York Botanical Garden provided copies of journal proceedings. My special thanks to Bob Goodman, Zhores Medvedev, and Matthew Meselson for reading the manuscript and for their valuable suggestions. Fred Chase was a brilliant copy editor.

Eleanor Randolph, as always, added her inimitable touches. I also thank Victoria Pringle for being a reader. Any mistakes are mine alone.

Index

Natural Sciences and Marxism
    (journal), 165
Nature (journal), 137
Navajos, 166
Nazarov, Abdun, 57
Needham, Joseph, 329*n7*
Nevitskaya, Princess, 18, 19
New Earth, The (Harwood), 97
New Economic Policy (NEP), 88,
    94–95, 101
New York Geographic Society, 311
New York Times, The, 213, 219, 226
Nicholas II, Tsar, 14–16, 24, 43, 46,
    52, 60, 91, 93, 151, 308
Nicholas, St., 17, 18
Nikonov, Alexander, 303
Nizhny-Novgorod Fair, 20, 141
NKVD, 229, 225–27, 235, 267, 278,
    288
    arrest of Vavilov by, 246, 249,
        330*n1*
    Chief Economic Department of,
        257
    interrogations by agents of,
        253–65
    Lubyanka prison of, 251
    Vavilov's death confirmed by,
        280–82
North Africa, botanical expedition to,
    108, 119–28
Novichkov, Alexei, 283

Odessa experimental station, 173,
    177–78, 200–201, 204, 229–31
Offerman, Carlos, 194
Ogonyok (magazine), 303
OGPU, 151, 154–57, 159, 181,
    186–89, 192, 193, 330*n1*, 332*n6*
    Economic Directorate of, 180, 194
Orbeli, Leon, 287
Order of Lenin, 205, 297

Origin of Species, The (Darwin), 103,
    319n11
Orthodox Church, 6, 7, 18, 45–46, 137

Pakistan, 113
Palestine, 118, 120, 121, 124, 304
Pamirs, 51–59, 61, 64, 92, 112, 180,
    308
Pangalo, Konstantin Ivanovich, 67
Pasteur Institute, 37, 186, 187, 188
Pavlova, Anna, 91
Pavlovna, Marina, 47
Peasant Labor Party (TKP), 156, 157,
    159, 192, 259, 331*n3*
Periodic table, 71, 72, 303
Persia (Iran), 54, 57, 64, 78, 119, 202
    seeds brought back from, 61, 92,
        115
    in World War I, 43, 44, 46, 47,
        49–52, 66, 308
Peru, 118, 122, 181, 184, 187, 268, 304,
    309
Peter the First (Peter the Great),
    Tsar, 233
Petrograd University, 319*n10*
Petrovskaya Agricultural Academy
    ("Petrovka"), 21–24, 27, 46, 49,
    53, 144, 190, 217, 250, 275, 307,
    319*n11*, 331*n3*
    arrests of professors at, 152, 156
    experimental farm of, 42
    female students at, 25, 30–31, 63
    history of, 24–25
    informers at, 157
    Lysenko at, 297
    Mendelian-Lamarckian debates
        at, 28
    Vavilov on staff of, 31–32, 34, 61
    vernalization research at, 134
Pioneers, 248
Pisarev, Viktor, 126, 128, 191

Illustration Credits

The photographs in this book come from three main sources: (1) Yuri Vavilov's private collection, which includes photographs published in his book *V Dolgum Poiski* [Long Search] (Moscow: FIAN, 2004); (2) Russian Academy of Sciences, *N. I. Vavilov Dokumenti, Fotografii* [Documents and Photographs], ed. N. Ya. Moskovchenko, Yu. A. Pyatnitsky, and G. A. Savina (St. Petersburg: Nauka, 1995); and (3) Archives of VIR, Vsesoyuzny Institut Rastenievodstva [All-Union Institute of Plant Industry, renamed for N. I. Vavilov]. Photographs from the VIR archives were also published in Igor Loskutov's *Vavilov and His Institute: A History of the World Collection of Plant Genetic Resources in Russia* (Rome: International Plant Genetic Resources Institute/IPGRI, 1999).

Photographs from these three main sources are credited below as follows: Yuri Vavilov; *Dokumenti, Fotografii;* and VIR archives and Loskutov. Original credits for photographs are included where available. For example, some photographs are sourced to Russian government archives RGIA (Russian Government History Archive, Moscow), ARGO (Archive of the Russian Geographical Society), and ARAN (Archive of the Russian Academy of Sciences).

Yuri Vavilov: *1, 17, 18, 21, 30, 33* (photo credit: Yuri Vavilov), *34, 35.*

Dokumenti, Fotografii: 2, 3, 4 (photo credits unavailable for these), 5 (RGIA, Moscow, f. 5, op. 3, d. 1051, l. 72–73), 7 (ARGO, f. 56, op. 1, d. 88, l. 1), 9 (ARAN, f. 1014, op. 5, d. 57, l. 1), *10* (ARAN, f. 803, op. 5, d. 153, l. 1), *11* (photo credit unavailable), *14* (ARGO f. 56, op. 4, d. 72), *16* (Fototeka VIR).

VIR archives and Loskutov: 6 (private archive of A. Kh. Bakhteev), *12, 13, 20, 22, 23, 24, 28, 29, 31, 32.*

Y. Rokityansky, *Academician Vavilov, Istoricheskaya Drama* [Historical Drama] (Moscow: Akademia, 2005): *8* (photo credit unavailable).

N. I. Vavilov, *Piat Kontinentov* (Moscow: Mysl, 1987), Eng. trans. *Five Continents*, ed. Semyon Reznik and Paul Stapleton (Rome: International Plant Genetic Resources Institute/IPGRI, 1997): *15* (photo credit unavailable).

E. N. Sinskaya, *Vospominaniia o N. I. Vavilov* [Recollections of N. I. Vavilov] (Kiev: Naukova Dumka, 1991): *19* (photo credit unavailable).
Valery Soyfer, *Lysenko and the Tragedy of Soviet Science* (New Brunswick, NJ: Rutgers University Press, 1994): *25* (orig. published in *Na Stroike MTS i Sovkhozov*, photo: Shaiket), *26* (photo credit unavailable), *27* (*Pravda*, January 3, 1936, No. 3 [6609], N. Kalashnikov and N. Kuleshov).

About the Author

Peter Pringle is the author or co-author of nine previous books, including the novel *Day of the Dandelion* and *Food, Inc.: Mendel to Monsanto—The Promises and Perils of the Biotech Harvest* (a *New York Times* Notable Book) and the bestselling *Those Are Real Bullets: Bloody Sunday, Derry, 1972.* The former Moscow bureau chief for *The Independent,* he has also written for *The New York Times, The Washington Post, The Atlantic, The New Republic,* and *The Nation.* He lives in New York City.

PLC 15